Native Americans on Network TV

FILM and HISTORY
Series Editor: Cynthia J. Miller

European Cinema after the Wall: Screening East–West Mobility, edited by Leen Engelen and Kris Van Heuckelom, 2014

Native Americans on Network TV: Stereotypes, Myths, and the "Good Indian," by Michael Ray FitzGerald, 2014

Bringing History to Life through Film: The Art of Cinematic Storytelling, edited by Kathryn Anne Morey, 2014

Native Americans on Network TV

Stereotypes, Myths, and the "Good Indian"

Michael Ray FitzGerald

ROWMAN & LITTLEFIELD
Lanham • Boulder • New York • Toronto • Plymouth, UK

Published by Rowman & Littlefield
4501 Forbes Boulevard, Suite 200, Lanham, Maryland 20706
www.rowman.com

10 Thornbury Road, Plymouth PL6 7PP, United Kingdom

British Library Cataloguing in Publication Information Available

Library of Congress Cataloging-in-Publication Data

FitzGerald, Michael Ray, 1953–
Native Americans on network TV : stereotypes, myths, and the "good Indian" / Michael Ray Fitz-Gerald.
pages cm. — (Film and history)
Includes bibliographical references and index.
ISBN 978-1-4422-2961-7 (cloth : alk. paper) — ISBN 978-1-4422-2962-4 (ebook) 1. Indians on television. 2. Stereotypes (Social psychology) on television. I. Title.
PN1992.8.I64F58 2013
791.45'652997—dc23
2013032834

∞™ The paper used in this publication meets the minimum requirements of American National Standard for Information Sciences Permanence of Paper for Printed Library Materials, ANSI/NISO Z39.48-1992.

Printed in the United States of America

This is the wish-fulfillment of the European colonist: his natural superiority voluntarily recognized [by his subject].

—Peter Hulme, *Colonial Encounters: Europe and the Native Caribbean*

We wonder what it is in the white American self that expects and demands unswerving love and loyalty from those it oppresses.

—Hernan Vera and Andrew Gordon, "Sincere Fictions of the White Self in the American Cinema: The Divided White Self in Civil War Films," *Classic Hollywood, Classic Whiteness*

Contents

Acknowledgments

I would like to thank Jonathan Bignell, my thesis chair, mentor, and adviser at the University of Reading (United Kingdom), not only for his crucial contributions but for his early and astute guidance in selecting this topic and helping me "cut to the chase." I also thank Simone Knox, whose valuable suggestions and attention to detail made this a much better work. I would also like to thank my MA adviser at the University of Florida, Bernell Tripp, for providing training and inspiration in historical methods and setting me on the right path.

Syed M. Khatib (f.k.a. Cedric C. Clark), whose work inspired me and with whom I consulted many times, was generous with his time and helped me clarify many of the ideas contained herein. Richard Allen of the Cherokee Nation provided an authentic American Indian view and set me straight on many of the issues discussed herein, particularly with regard to Cherokee culture. Leighton C. Peterson of Miami (Ohio) University provided advice on Navajo culture. Generous comments and expenditures of time and moral support were provided by reader/reviewer M. Annette Jaimes of San Francisco State College. Peter Rollins graciously provided footage of an extremely hard-to-find documentary. Valoise Armstrong at the Eisenhower Library in Abilene, Kansas, was helpful in providing crucial documents, as was Amy Reytar at the National Archives in Washington, D.C., and John Gibson at the Detroit Public Library.

The person most responsible for this book is Cynthia Miller, series editor at Rowman & Littlefield. I also thank R&L's senior editor Stephen Ryan. I must also thank Susan Allen for her patience and moral support while I spent several years researching and writing.

I dedicate this book to the memory of my mother, who was a television buff ever since I can remember. While researching my doctoral thesis, I would call her at all hours to ask her questions about TV shows and movies, and she was usually able to provide a reliable answer, most of the time off the top of her head. She took great pride in being able to contribute to my academic pursuits even though she herself had little formal education.

Introduction

Origins and Applications of Stereotypes

> *A pattern of stereotypes . . . is the projection upon the world of our own sense of our own value, our own position and our own rights. . . . They are the fortress of our tradition, and behind its defenses we can continue to feel ourselves safe in the position we occupy.*—Walter Lippmann [1]

> *No identity can ever exist by itself, without an array of opposites, negatives, oppositions. Greeks always require barbarians.*—Edward Said [2]

Whether as a villain, subservient friend, or a hapless victim of progress, the American Indian has been a crucial figure in U.S. literature, film, and television.[3] Many Euro-Americans have rarely if ever met a real American Indian; most ideas they have about them come from film and television.[4] Even when they are set in the contemporary era, many if not most Indian stereotypes are derived from or descended from stereotypes emanating in European colonial discourse; some of these are literally hundreds of years old and survive in updated, modified, or recombinant forms. Some have not changed substantially since Pocahontas or Friday. U.S. literature, film, and television tend to purvey rehashed tropes based on figments of the colonial imagination, some—such as Pocahontas—going back as far as the earliest English settlements, some as old as the Spanish conquest.

Stereotypes of American Indians have often been used to justify their repression and extirpation—both literal and symbolic—since the era of the Pilgrim settlers. Paradoxically, even as many English-speaking settlers feared and loathed American Indians, a great many also envied them and wanted to be like them.[5] Presumed hypermasculine qualities of Native Americans were (mis)appropriated and assigned to Anglo-American heroes (mostly Indianized white men) in order to distinguish supposedly indefatigable Anglo-Americans from effete Europeans.

This study examines how images of American Indians were (mis)represented, transformed, and distorted to fit whatever political or cultural concepts that may be useful for the dominant elements in U.S. society in any given era. What is amazing is how durable—and malleable—these images have been over the course of, literally, centuries. Such stereotypes have varied according to each era's ideological necessities: whatever Euro-Americans needed Indians to be, they made them that thing. What is more, stereotypes "help white America contain and numb the reality of past guilt and present injustice."[6] Daniel Francis has posited that the continued use of outdated stereotypes is a response to "a deep anxiety that non-natives have about our place in North America and a deep need to legitimate our presence here. . . . Movies project onto Indian characters the uncertainty non-natives feel about the justice of our history and our right to occupy the land."[7]

As several scholars have shown, the American Indian stereotype has two poles with little or nothing in between.[8] The Indian has usually been depicted in literature, film, and television either as an evil, murderous enemy or—often simultaneously—as an innocent, sometimes helpless child of nature. Indians are rarely seen as "regular" people. Whether Bloodthirsty Savage or Noble Savage, in either case they are savages, less than human. Like children, they are incapable of reason and so can only be dealt with as objects to be *managed*, either through force or acts of benevolence. In either case, they have little or no agency of their own.

A Good Indian was one who helped the white man in his quest to dominate the land. This trope probably originated with the myth of Pocahontas, who supposedly found John Smith so irresistible that she sneaked away from her people to warn him of their impending attack. To put this in perspective: had a U.S. citizen done anything like this during an international conflict, it would almost certainly have been regarded as treason. Indeed, Leslie Fiedler calls her "our first celebrated traitor to her own race."[9] Pocahontas became the first in a long line of American Indian stereotypes offering what I call *endorsement* of Euro-American domination. This Indian-collaborator myth might more accurately be seen as a fantasy or a wish fulfillment invented by Euro-Americans to (retroactively) legitimate and justify their domination of the continent. Literary critic Diane Krumrey states that "[Euro-]Americans need to retell the story of the first contact between English settlers and Native Americans to make it a loving, interactive and mutually satisfying encounter, to knit themselves into the landscape."[10] Fiedler says much the same, calling the Pocahontas story "a myth of reconciliation between the races."[11]

Another primordial example is the subservient helpmate or "ethnic sidekick," probably derived from Daniel Defoe's Carib character Friday in *Robinson Crusoe*.[12] There are competing groups of natives in *Crusoe*, and it is not clear in Defoe's story which group Friday belongs to. It is commonly assumed that Friday was an Arawak while his tormentors—from whom Crusoe saves him—were Caribs; in any case, the difference between the two was probably artificial, superimposed by European outsiders based on erroneous assumptions. The most likely explanation is that the fierce Caribs were themselves a subset of Arawaks, but in the minds of the Spanish, "Carib" became synonymous with "cannibal," while Arawaks were described as gentle and generous to a fault; to the Spanish mind, these must have been two entirely different peoples, hence the traditional division of the natives into "friendlies" (Arawaks) and "hostiles" (Caribs).[13] This Good Indian–Bad Indian duality reappears throughout Euro-American (colonialist) literature under such binaries as Noble Savage/Bloodthirsty Savage and so on.

Ultimately, the Good Indian is one who is astute enough to recognize the white man's natural superiority and acquiesces to his control. This has been a remarkably resilient cliché that reappears in many guises, under many circumstances—serving various cultural and political demands—for centuries. As this study will demonstrate, this particular Good Indian figure has been the prevailing American Indian representation on U.S. television. This is primarily because nationwide network television appeared *after* World War II, when depictions of Bad Indians had already become a political liability. The Good Indian appears as an enforcer of the dominant group's norms and is sometimes shown as an active collaborator with Anglo-American military forces as on such programs as *Brave Eagle* (1955), *Broken Arrow* (1956–1958), and *Daniel Boone* (1964–1970). Several series, such as *Law of the Plainsman* (1959–1960), *Hawk* (1966), *Nakia* (1974), and *Walker, Texas Ranger* (1993–2001) go even further, depicting the Indian character becoming almost entirely Anglo-Americanized. However, these characters retain enough "Indianness" to make them remarkably intuitive as well as both spiritually and physically superior to whites.

Whether bad or good, American Indians have mostly been portrayed as natural, land-based *obstructions* that stood in the way of growth and progress—that is, the expansion of global, neoliberal capitalism—and like trees, required removal. This "doomed-victim" trope—a variation of the Noble Savage—is one of seven common stereotypes identified by scholar Mary Alice Money.[14] Another form of stereotyping was the elimination of Native Americans altogether, a lie by omission that implies that they simply do not—or perhaps should not—exist. This suggests an imaginary—often all-white—world where certain people simply do not appear.[15] George Gerbner and Larry Gross coined the term "symbolic annihilation" to describe this elimination of minorities and women from television.[16] However, this concept had already been recognized by Gerbner's onetime research assistant, Cedric C. Clark, in 1969: Clark termed such conspicuous absence "non-recognition."[17] Gerbner and Gross's term goes further, however, implying a sort of metaphorical genocide.

PURPOSE OF STUDY

Television representations are the most recent stage in a four-hundred-year history of politically and ideologically tinged representations from Puritan "captivity narratives" to western dime novels to films and finally to the airwaves (and cable). This study addresses representations of American Indians on network television, what these meant at certain historical moments, what they mean in a larger context of a historical process that has evolved over hundreds of years, and how they were used in relation to the dominant ideology. There are some interesting, identifiable, and persistent patterns among these representations when considered as a body of work. Most studies dealing with television representations have been collapsed into books on filmic or literary representations. What is more, few if any studies of televisual representations offer detailed analyses of the ways in which the images themselves—their presentation, framing, shot angles, mise-en-scène, and so on—articulate racial ideologies.[18]

"Westerns"—that is, programs set in the western region of North America during the Euro-American conquest—contain the bulk of American Indian televisual representations, but these common—and conveniently malleable—depictions also reappear in "crime" series, especially ones set in the contemporary West; in addition, there are programs that combine both (or more) genres. Television's role in the military-industrial complex and its nationalist function during the Cold War play a large part in myth-ifying—and mystifying—the conquest of Native America, making the United States appear naturally good and just. American Indian characters also served in roles that were related to more contemporary issues of crime, poverty, racism, citizenship, and Anglo-American (i.e., white) masculinity. This study examines these aspects of American Indian representations, including the ways the western genre's visual tropes and narrative strategies serve to perpetuate unspoken, unquestioned ideological assumptions—in Roland Barthes's term, *myth*.[19] Collectively, westerns comprise Anglo-America's "foundation myth."

This study deals with no less than five intertwined histories. It would be difficult to contextualize or gain any deep understanding of any one of these without having at least a passing knowledge of the others:

1. The history of American Indians themselves (of which I admittedly deal with only superficially).
2. The history of the conquest of the North American continent by Europeans and their descendants, including the period of the Puritan "set-

tlement" and the period colloquially referred to as "the winning of the West," along with their accompanying narratives.[20]

3. The history of World War II and the subsequent Cold War and how images and ideologies from "the winning of the West" were resurrected and *reapplied* to suit the ideologies of these periods.
4. The history of U.S. network television and its roots in the defense industries as well as how these are imbricated in the Cold War and what is referred to in political-science circles as "the national-security state."
5. The history of the African American civil rights movement, which reached critical mass during and after World War II.

It is imperative to address these histories individually but more so to consider how they interact. I have not closely examined the history of American Indians themselves because one could not do justice to this topic in the space of this study, yet of necessity there are elements of this history contained herein.[21] I am more concerned here with deconstructing *representations* of Indians and what they mean to Anglo-Americans as well as *why* they mean what they do.

PREVIEW OF CHAPTERS

This introduction provides a survey of the field of media representation, including a review of the relevant literature, along with an explanation of research aims, questions, and methods used in this study. Chapter 1 outlines and explains the results of several years' worth of research and introduces the framework for this study, which is largely based on a 1969 *Television Quarterly* article by communications scholar Cedric C. Clark (now Syed M. Khatib). Clark's Evolutionary Stages of Minorities in the Mass Media (the name given his framework by scholar Alice A. Tait) analyzes the ways minority characters move from nonrecognition (symbolic annihilation) to ridicule to finally gaining respect.

The stage Clark seemed to focus most on is what he calls Regulation, in which *subaltern* characters (i.e., members of subordinate groups or "second-class citizens") become enforcers of the dominant group's norms.[22] Though Clark was mostly concerned with African American representations, I found his framework to be extremely predictive when applied to Native American characters: his "Regulator" is closely related to the tropes of Pocahontas, who purportedly "finked" on her father, as well as Friday, a Carib (or an Arawak) who served as Crusoe's "Indian companion." These tropes—consisting of natives who astutely recognize the purported superiority of European culture (and technology)—not only survive, they thrive. Using quantitative methods, I found that the more prominent the Indian character is in any given program, the more likely he (rarely if ever a she) is to be depicted as a Regulator or enforcer. After exhaustively identifying all possible prospects, I examined the most prominent characters (in terms of narrative centrality) and found all of them to be, as Clark would predict, Regulators. The next step was to focus on these by employing in-depth analysis, as there are ambiguities that need to be recognized. Moreover, not all recurring American Indian characters conformed to this category, only the ones who were featured in central or "starring" roles. The most notable exception to Clark's Regulator rule pertains to the 1990–1995 CBS series *Northern Exposure*, in which *none* of the characters fits into his Regulator category. What is more, this program rarely relied on stereotypes, allowing Alaska natives to be regular people with regular problems and their own idiosyncratic quirks. This is unusual for U.S. television, which relies heavily on stereotypes.[23] Thus *Northern Exposure* was the exception that proves the rule. Despite this, *Northern Exposure* still resembles colonialist "pioneer" adventures in that Alaska serves a similar function as the exotic East: it serves as "an isolating element, as sexually and morally separate imaginary territory" where anything can happen.[24]

Chapter 2 examines the seminal television western *The Lone Ranger* and supporting character Tonto and his role as the Ranger's man Friday. Tonto is indeed a Regulator: he helps the Ranger in his vigilante efforts to promulgate Anglo-American law and order—and to further the expansion of Anglo-American capital.[25] Chapter 2 closely examines this Good Indian trope and notes its many historical manifestations as well as how Tonto became a sort of template for future representations in this mode. *The Lone Ranger* purveys what media sociologist Herman Gray calls an *assimilationist* discursive strategy, in which the American Indian character more or less renounces his own identity in favor of mainstream (i.e., white) culture.[26] Moreover, *The Lone Ranger* series contained political and foreign-policy implications, even though it was targeted at children.

Chapter 3 examines the television version of *Broken Arrow*, adapted from a 1950 Delmer Daves film, in which the Apache character, Cochise (loosely based on Cheis, a real-life Apache leader), is ostensibly presented as equal to the white protagonist. This program evinces what Gray calls a *pluralist* discursive strategy in which the two groups, white and Indian, are seen as "separate but equal."[27] Cochise is indeed a Regulator, but unlike Tonto, he possesses near-superhuman qualities (some of which have traditionally been associated with Native Americans in literature).[28] This Anglicized-Indian figure seems to be moving toward a fusion comprising "the best of both worlds" as does the white character, Tom Jeffords, who becomes Indianized. This development, in which both figures "meet in the middle" and take on each other's characteristics, had been evolving in literature since Cooper's

Natty Bumppo and Chingachgook in a series of novels called the "Leatherstocking Tales" (1827–1841).

This hybridity nears completion with character Sam Buckhart, protagonist of *Law of the Plainsman*, examined in chapter 4. This story, like *The Lone Ranger*'s, also represents an *assimilationist* strategy in which the minority character becomes part of the white world, leaving his own behind. Also in this chapter I note that the program's production company, Twentieth Century-Fox, eagerly collaborated with the Eisenhower administration to provide positive images for public consumption (or "propaganda" in Darryl F. Zanuck's own words). All this occurred during the federal government's implementation of a policy dubbed "termination," in which treaties with sovereign nations were unilaterally terminated or abrogated. Chapter 4 also discusses representations (and strictures) of masculinity. Can there be any doubt that the western genre is primarily concerned with issues of masculinity?[29] Representations of white and nonwhite masculinities differ drastically; we see this in the contrasting ways in which Indian Agent Tom Jeffords and Apache leader Cochise are depicted.[30]

Chapter 5 examines the NBC series Daniel Boone (1964–1970), which featured Cherokee Chief Mingo, the "half-breed" son of the governor of the Virginia colony. Mingo serves very much in the mold of Friday to Daniel Boone's Crusoe. What is more, Mingo is a prototype hybrid, the embodiment of the best of both races; he has a great deal of European "cultural capital"—including an Oxford education—and hence can "pass" as white anytime he pleases. Identity for him is a matter of choice.

Chapter 6 examines the 1966 crime drama *Hawk*, in which the Indian character is reinvented as a modern playboy (à la Hugh Hefner) as well as awarded attributes of Ian Fleming's superspy James Bond. Here we see the final stages of an *assimilationist* strategy in which the protagonist's Indianness is subsumed, nearly invisible (the same holds true for the character Chief in *Garrision's Gorillas*, 1967).

Chapter 7 looks at the 1974 police drama *Nakia*, which moves back toward what Gray calls a *multicultural* strategy.[31] It seems the assimilationist strategy had gone too far, having swallowed up the identity of minorities; the new paradigm emerging in the 1970s was (and still is) one of celebrating difference or "diversity." Deputy Nakia Parker's Navajo origins are awarded a degree of respect, and he, at first, questions his own loyalties. Ultimately, however, he is recuperated into white-male hierarchy.

Chapter 8 examines the ultimate hybrid, Cordell Walker, a half-Cherokee lawman in the "contemporary western" *Walker, Texas Ranger* (1993–2001). As a part-American Indian, his absorption into and unquestioning loyalty (as well as usefulness) to the Anglo-American way of life is complete—yet he draws upon his Indianness whenever it suits his purposes. Walker combines the best characteristics of the Anglo-American lawman *and* the Native

American; he is both cowboy *and* Indian, Lone Ranger *and* Tonto. Thus these two cultures, Anglo-American and American Indian, stand together as one—or so the program's producers (which included star Chuck Norris, who claims Cherokee ancestry) might have liked viewers to assume. Much like *Superman*'s Clark Kent, Walker has a "secret identity": he can be either white or Indian, as suits his needs at any given moment.

The conclusion in chapter 9 explains how resilient this Pocahontas-style character—and its many variations—has been throughout the history of the Anglo-American conquest. It also offers explanations for the necessity of creating such characters and how these operate on an ideological level, providing satisfaction for certain viewers and dealing with *cognitive dissonance* (i.e., discomfort) created by the contradictions between the United States' self-proclaimed image as a land of "liberty and justice *for all*" and the ways minorities have historically been treated. Daniel Chirot and Clark McCauley define dissonance as "an unpleasant arousal that comes from seeing ourselves as having chosen to do something wrong, stupid, or sleazy—that is, something at odds with our positive self-image."[32] This dissonance was especially severe during the Cold War, since Russian and French newspapers were quick to point out such contradictions.

METHODS

As an academic discipline, the communications field, especially the areas of public relations and advertising, largely serves to train media technicians in the latest opinion-molding techniques—in other words, how to manufacture more effective manipulation. In order to "see through" such sophisticated, scientific techniques, it is necessary to go beyond communications studies to a multidisciplinary approach. Scholars interested in more responsible media must look to various disciplines in order to gain a broader perspective and synthesize a more thorough inquiry. Even the relatively few critical communications scholars who ask difficult questions have been decades behind literary analysts and cultural studies scholars in implementing these interrogations. For one thing, quantitative social-science methods and statistics cannot address the quality or the meaning of minority representations. Theories from the fields of literary criticism, semiology, linguistics, sociology (especially symbolic interactionism), anthropology, psychology, even psychoanalysis, should be considered. This study will employ a multidisciplinary approach; its methods include quantitative, qualitative, historical, institutional, political economy, and critical analyses.[33]

This study began as a listing of every American Indian character that has regularly appeared on U.S. network television from 1949 to 2009, garnered from various sources including the Internet Movie Database, TV.com, TV

Acres, and encyclopedic references such as McNeil's *Total Television* and Brooks and Marsh's *Complete Directory to Prime Time Network and Cable TV Shows* along with various books and articles mentioning Native American television characters.[34] These were entered into a spreadsheet so that they might be sorted by year, network, settings, gender, and occupations of the characters, as well as their prominence in the programs. This helped crystallize certain themes and motifs, such as how all the most prominent (i.e., central) characters fit Clark's Regulation stage. Having exhausted the limits of quantitative work, I ventured into qualitative methods, particularly textual and visual analyses, in order to examine what these images and themes appear to signify, both in contemporary terms as well as in their historical and traditional contexts. Although I often relied on secondary sources such as literary histories for overviews, I obtained many primary documents, including copies of the programs themselves, along with relevant documents where necessary and feasible.

Some programs were deliberately excluded; these include miniseries, which did not even materialize on television until *Roots* in 1977; made-for-TV movies, another relative latecomer; and local and regional programs. The present study focused exclusively on weekly programs under the assumption that regular, recurring series have more ongoing impact on U.S. cultural consciousness than miniseries and made-for-TV movies. The viewing public is aware that the life span of these is but a flicker. I focused on network rather than syndicated programs for similar reasons: syndicated programs are sold on a market-to-market basis and usually do not enjoy the broad, nationwide penetration network programs entail. The decision to exclude what I feel are marginal items was based on Gerbner's cultivation theory, which posits that the more regularly images are seen by viewers, the more plausibility viewers assign to them and the more likely they are to be accepted as "normal." Stuart Hall basically makes the same claim.[35] Aumont posits a similar theory regarding filmic conventions and stereotypes, which he dubs "plausibility."[36]

THEORIES OF REPRESENTATION—AND WHY IT MATTERS

As literary critic and cultural theorist Edward Said explained, "Institutional forces in modern western societies . . . shape and set limits on the representation of what are considered essentially subordinate beings; thus representation itself has been characterized as keeping subordinate subordinate, the inferior inferior."[37] There are many theories of how representations work, but few address who benefits and who is harmed. Such theories are, at least in the United States, marginalized: for example, in the communications field, these studies are usually lumped—or ghettoized—under the rubric "critical/cultural studies." Most theories of representation that address underlying issues of who benefits and who is harmed are found *outside* of the field of communications. Not only is the present study multidisciplinary, it is transcontinental in perspective—undertaken under the auspices of Drs. Jonathan Bignell and Simone Knox of the University of Reading (United Kingdom) and thoroughly influenced by the ideas of Stuart Hall, one of the founding fathers of British cultural studies.

REVIEW AND SUMMARY OF THE RELEVANT LITERATURE

Though largely discredited as a political scientist, Karl Marx has been canonized as a founding father of sociology; his view of the clashing interests between owners of industry and workers has been incorporated into what U.S. sociologists call "conflict perspective." In media and cultural studies, his views survive in the perspective called "political economy," which looks at the owners of the "culture industries" and whether the viewpoints expressed by the media reinforce or normalize their owners' interests. Marx used the term "ideology" to denote the ways in which the bourgeoisie (in this case media owner and managers) seek to normalize their interests by making these appear natural and desirable. In other words, elites seek to entice the lower classes into agreeing that what is best for them is best for all. This is often an inherently ethnocentric view in that usually only the interests of the dominant group are recognized. What is more, these values are very often *mythicized*, that is, associated with archetypal concepts, and thus rarely questioned.[38] However, cultural-studies scholars found many of Marx's ideas, although groundbreaking, to be insufficient and overly reductive. For example, Marx failed to take into account interests other than class, such as gender, ethnicity, and so on.[39] In the 1930s, Antonio Gramsci addressed some of the failures of Marx's theories and in the process took the concept of ideology a step further. Gramsci concurred that media owners indeed constituted a distinct class of manufacturers, and their products should reflect their class interests and norms. However, Gramsci recognized that this process does not have to be conscious nor deliberate and called it "cultural hegemony."[40] Media sociologist Todd Gitlin often invokes Gramsci's term "hegemony."[41] U.S. social critics Noam Chomsky and Edward S. Herman posit a remarkably similar theory, calling it "manufacturing consent," a modified (and ironic) variation of public-relations specialist Edward Bernays's *The Engineering of Consent* (1969).[42] In the 1930s, the Frankfurt School "critical theorists," including Theodor Adorno and Herbert Marcuse, began asking similar questions, and their ideas have reverberated throughout the modern era.[43] They were among the first to criticize Marx's assumptions and move toward a more dialectical framework, which is that the media do not simply function as a vehicle for the propagation of the elite's interests; rather, mass culture became of necessity—in order to make money—a site of negotiation between manufacturers (the "culture industries") and consumers. This is Hall's position as well. French neo-Marxist Louis Althusser, however, leaned back toward the top-down model, noting that the media generally behave as an arm of the state (or as an Ideological State Apparatus) regardless of whether they are state or privately owned.[44] Althussser's concept, however, requires an expanded definition of the term "state": the state here means not only the government but must include a group of private corporations that exert influence over both government and media. Thus the media, the government, the telecommunications corporations and electrical manufacturers who contract with the government, and many of television's largest advertisers share similar interests; therefore they have every incentive to cooperate as a bloc. Althusser's ideas have been echoed by media critic Robert McChesney.[45]

In the 1960s, media sociologist George Gerbner, dean of the Annenberg School of Communications at University of Pennsylvania, formulated a theory that explained Marx's idea of normalization and how it probably works: through *repetition* and emphasis of certain representations, to the exclusion of others. In other words, the more a viewer sees something, the more believable it becomes. This also has a cumulative effect of making the repetitive appear normal and the unusual weird or abnormal; Gerbner called this *cultivation*, a process that applies to television in a similar fashion as Aumont's theory of *plausibility* does to film.[46] Also in the 1960s (and beyond) British cultural studies—the Birmingham School under sociologists Richard Hoggart and Stuart Hall—sought to rectify the problem that Marxian analysts viewed ideology exclusively from a class-based perspective, ignoring race and gender.[47] Another key issue Hall et al. recognized is that media do not simply reflect the ideology of their owners but comprise an arena in which meaning is contested, synthesized, negotiated, and tentatively agreed upon.[48] Hall and Gerbner are actually quite close on the matter of cultivation: Hall writes that the cultural industries do not have the power to unilaterally impose their values on a populace, but they do have the power, "*by repetition and selection*, to impose and implant such definitions of ourselves as fit more easily the descriptions of the dominant or preferred culture."[49] This synthesis is conducted on both an individual level (by viewers themselves) as well as on an industrial level: viewers can only make meaning of images that are made available to them. Thus both parties and their contributions are in play simultaneously.[50] Granting that the media-owning elite enjoys a preponderance of power in creating meaning, their power is not absolute and is often overdetermined, but to what extent is still being debated. In other words, media owners *do* control the representations and the stories but not which of these consumers select. Moreover, the more they give consumers what they want, the more "successful" the owners will be. Without high ratings and consequent advertising revenues, a U.S. network might not survive. Thus, *contra* Marx, ideology is not a primary concern, although in Zanuck's case, for example, it was a serious secondary concern. Making money and staying in business is the primary concern, at least for U.S. networks and producers. The revival of Gramsci's paradigm in the 1980s stemmed from the fact that he was able to synthesize both perspectives, thereby resolving a debate that had been (and still is) flaring up in communications and other fields dealing with public discourse. Ultimately the theorists have been moving from an Orwellian (top-down) model, in which ideology is imposed from above, to a more Huxleyan proposition, in which both parties (media owners and public) willingly participate.[51] Neither party can be deemed innocent. This is the problem with the traditional Marxian approach: it presumes consumers are innocent "cultural dupes." They are not. Oftentimes media elites and propagandists simply give consumers virtual permission to do what they already want to do—scapegoat minorities and foreigners, for example—by making this seem normal, unremarkable.

STEREOTYPES, DEMONIZATIONS, AND TOTALIZING TROPES

The term "stereotype" should be defined and examined. In printing terms, a stereotype is a plate consisting of an entire page with its various components already integrated. In 1922, journalist Walter Lippmann, who had served on President Woodrow Wilson's Committee for Public Information during World War I, borrowed the term as a metaphor to describe oversimplified, totalizing representations of ethnic groups.[52] Communications science was also launched during this post–World War I period, as sociologists Harold Lasswell and Paul Lazarsfeld brought quantitative methods to studies of propaganda in an effort to understand why and how propaganda had worked to mobilize the U.S. populace.[53] Much of the war propaganda had been based on demonization of Germans. As Glen Ford, a columnist for *Black Agenda Report*, points out, "The power to demonize is a weapon of war."[54] After the impressive success of propaganda and "psychological warfare" *on all sides* during World War II, communications studies became a lucrative research field. Scholars, aided by U.S. government grants, examined how, why, and to what extent such tactics worked.[55] However, modern communications scholars have mostly failed to ask what the meanings of stereotypes are, what their political purposes might be, and—the key question—whom they benefit. As Richard Dyer notes, the problem with stereotypes is not that they are intrinsically wrongheaded; what is crucial is "who controls them and defines them [and] what interests they serve."[56]

American Indian stereotypes are "part and parcel" of what Ward Churchill calls "colonizer discourse," used by dominant groups to denigrate other peoples, justifying their subjugation and the seizing of their resources.[57]

Such incursions must be rationalized even if this means using the most distorted contortions of logic, because to face them would expose the discrepancies between the noble beliefs colonizers harbor about themselves and the ways their own behavior blatantly contradicts their beliefs. The more obvious the discrepancy, the more elaborate the rationale must be, often involving a great deal of time and (psychic) energy. One early explanation of the rationalization function involved in stereotyping came from Charles de Secondat, baron of Montesquieu, who wrote in 1748, "It is impossible for us to suppose these creatures to be men, because, allowing them to be men, a suspicion would follow that we ourselves are not Christians."[58] Montesquieu refers here to African slaves, but this could just as easily apply to any subaltern group: it is necessary to avoid recognition of these creatures as human in order that we may persecute them without incurring guilt. Montesquieu's observation demonstrates why a culture needs to "paper over" its all-too-obvious hypocrisy with a myth or ideology. In psychological terms, society employs ideology to deal with *cognitive dissonance* caused by awareness of contradictions between its values and its behavior.[59] Montesquieu thus unmasks the (false) proposition that only white Europeans were entitled to Christian charity.

In the wake of World War II and the struggle of former colonial states for independence, anticolonialist and Africanist authors began articulating propaganda with colonizer discourse. One of the most influential of these was Frantz Fanon, a French-speaking psychiatrist from Martinique, whose views had largely been inspired by his teacher, "pan-Africanist" theorist and poet Aime Cesaire, also from Martinique.[60] In 1952, Fanon began publishing investigations into the intentions and purposes behind ethnic stereotypes and concluded that these were mostly designed for the benefit of colonizers who felt guilty about what they were doing in the colonies.[61] Fanon's psychoanalytical approach—he felt a society could, much like a person, be psychoanalyzed through its rhetoric—pioneered what would later be called "postcolonial studies." In 1958, Tunisian-French anticolonialist Albert Memmi investigated the purposes of derogatory depictions of North Africans, drawing much the same conclusions (1965).

Also in 1957, French semiologist Roland Barthes analyzed a cover of *Paris Match* magazine (figure 0.1) published during the period of France's imperial struggles in Africa and Algeria, not long after its expulsion from Vietnam. Barthes writes,

> I see very well what [this picture] signifies: that France is a great empire, that all her sons, without any color discrimination, serve faithfully under her flag, and *there is no better answer to the detractors of an alleged colonialism than the zeal shown by this Negro in serving his so-called oppressors.*[62]

Figure 0.1. Cover of *Paris Match*, June 25, 1955.

This remark cuts through to the heart of the matter and answers the question of why there are so many Regulators. According to Barthes, the native enforcer of the dominant group's norms (the Regulator) operates in the realm of *myth*. A myth, in Barthes's terms, is a dialectical invention (a synthesis of two poles) that bridges opposing concepts; it miraculously reconciles mutually exclusive propositions (this is very similar to Marx's use of the term "ideology"). *Myth*, Barthes writes, "finds a third way out" of a logical impasse; "it gets out of a tight spot thanks to a compromise."[63] In this case, the black soldier *mediates* between the two poles, one being French imperialism (the colonizers), the other its annexed peoples (the colonized). The way out of this logical dilemma, then, is the soothing myth that the *colonized welcome their own subjugation* and, what is more, *are willing to fight to expand the empire's reach onto other peoples.* French imperialism is deemed desirable for all concerned. This representation, the subaltern imperial enforcer (in British India, they were called sepoys), presents this state of being as implicitly normal, unremarkable. Barthes adds, "[D]riven to having either to unveil or liquidate the concept, [myth] will *naturalize* it."[64] Thus this trope's underlying assumption becomes an article of faith, never questioned, for if such assumptions *were* scrutinized, they would fall apart.

Stereotypes Specific to the United States

The image of the United States as a land of liberty and justice for all has historically been marred by its treatment of minorities, most notably African Americans, but also of other nonwhite minorities—even at some points including certain groups of whites, such as Irish Catholics, Italians, Poles, and so on. In 1920, black leader W. E. B. DuBois noted that, for centuries, Europeans had "exhausted every ingenuity of trick, ridicule, and caricature on black folk."[65] In the mid-1950s, novelist and social critic James Baldwin delivered insightful analyses into the psychological purposes behind film and television stereotypes of African Americans and how these *comforted* white audiences by pandering to their prejudices.[66] Both DuBois's and Baldwin's observations correspond with Clark's first stage, Ridicule.

Addressing American Indian stereotypes, Leslie Fiedler used literary anthropology to examine these as justifications for subjugation and extirpation as well as the various ways Euro-Americans (mis)appropriated the identities of American Indians as their own.[67] This was akin to Fanon's psychoanalyzing the French colonizers' rhetoric. Richard Drinnon's trilogy of historical and psychoanalytical analyses of American literature demonstrate how devices such as stereotypes, demonization, and other psychological "defense mechanisms" were deployed and recycled to serve objectives of conquest and colonization from Puritan times to the Spanish-American War and later to Vietnam.[68] Political scientist Michael Rogin employed a Lacanian psychoanalytical approach in examining these techniques, especially those used in western movies.[69] Ward Churchill melded communications studies with postcolonial theory, history, law, and other disciplines with specific regard to American Indian representations in a study aptly titled *Fantasies of the Master Race.*[70]

Noted film scholar Richard Dyer demonstrated that people of European descent seem to particularly enjoy seeing themselves depicted in godlike terms. In his fascinating study of western art, *White*, Dyer demonstrates the ways various media are exploited by people of European descent to represent themselves as divine.[71] There are numerous visual clichés that connote the godliness of the white race. Anglo-Americans in particular tend to depict themselves as God's "chosen people," destined to lead the world, a theme that stems from the country's Puritan origins, a tradition that still survives, as this study will reveal by identifying some of these tropes and unveiling their underlying assumptions.[72]

As Lippmann points out, stereotypes usually contain some underlying ideology that places the user above the "object" being signified; in other words, stereotypes presume the natural domination of one group over others: the unspoken assumption such caricatures rely upon is that the people they represent are somehow deficient, inferior.[73] In other words most people enjoy stereotypes because *they make us feel better about ourselves* in comparison to those "other" folks. Structurally, stereotypes only make sense when compared to one another. A negative stereotype must have a positive antipode: for example, the Good Negro or Uncle Tom character from Harriet Beecher Stowe's *Uncle Tom's Cabin*, published in 1852, serves as a counterpart to the Bad Negro trope such as the venal, sexually depraved black men depicted in D. W. Griffith's *Birth of a Nation* (1915).[74] Other examples of simultaneous good/bad stereotyping are the Noble Savage and Bloodthirsty Savage Indian stereotypes. Neither exhibits a full range of human characteristics, for these are not representations of actual people, nor do they under the slightest scrutiny resemble real people. *The fact that they do not even remotely resemble real people reveals these figures to be not "representations" at all but rhetorical devices—metonyms—designed to provoke an automatic reaction*. Rhetorical devices are designed for a purpose: to persuade. Moreover, these operate on the level of emotion (pathos), not logic (logos).

The Noble Savage trope has been in use since the seventeenth century, when English poet John Dryden used it in his 1687 play *The Conquest of Granada by the Spaniards*. Positive stereotypes of Native Americans were invoked "as a polemical device to criticize European institutions."[75] French settlers and missionaries were more apt to paint a positive picture of Native Americans than were the English. The difference between the English and French views might be explained by the fact that England had already undergone a revolution by the time its colonizers began arriving in North America, whereas in France revolution was still in the offing. French reformers appropriated the image of the Noble Savage as a symbol of a more democratic, less authoritarian way of life (American Indians indeed enjoyed less hierarchical societies).[76] However, at the same time, French conservatives used similar images to illustrate the brutish existence and degradation that would supposedly befall France if her established institutions were to founder.[77] Thus began the appropriation of images of American Indians as political puppets, used for purposes with which they themselves had no connection, a tradition that continued well into the 1990s, as this study will demonstrate.

Negative Stereotypes and Their Destructive Influence

The use of negative stereotypes is often conflated with what is commonly called "racism"; however, negative stereotypes can be employed against people of the *same* race, ethnicity, or nation. For example, Hindus and Muslims in India, or Hutus and Tutsis in Rwanda, belong to the same race; therefore prejudice here cannot correctly be termed "racism." What is really at issue in these cases is the dominance of one group—or one class—over another.

Dehumanizing is an extreme form of stereotyping. Extermination campaigns throughout history have been accompanied if not activated by campaigns of hatred fueled by demonizing stereotypes. Calling another group "evil" can be a vague but effective justification for attacking its members. This too entails a structural differentiation:

> If we call someone else "evil," what we mean is that *we* are the good people. And if we are good and they are evil, then we are not only off the hook for negotiating with them, but it becomes our "moral" duty to exorcise them, to exterminate them and make the world safe.[78]

One of the earliest written examples of the uses of negative stereotypes for such purposes appears in the Old Testament wherein the Israelites demonized the Canaanites as blasphemers and idolaters who were to be exterminated so that their land could be expropriated.[79] The Spanish conquistadores used stereotypes to dehumanize natives in South and Central America and justify their conquest and extermination. In the words of Juan Gines de Sepulveda, "These people are as inferior to Spaniards as children are to adults and women to men."[80] Francisco de Vitoria wrote, "Although these barbarians are not altogether mad, they are not far from being so. They are not capable of governing themselves any more than madmen or wild beasts."[81] The conquistadores used these justifications to liquidate an estimated 60 to 80 million aboriginals in the Americas.[82] As Bhabha writes, negative stereotypes are generally part of a campaign to justify conquest.[83] In North America, the Mattaponi (the group Pocahontas belonged to) authors of *The True Story of Pocahontas* provided an example of the ideology inherent in the use of the term "savage":

> Because the English [settlers of Jamestown] considered the Powhatan people savages, they considered it okay to kill them and take their land. There has to be some kind of conviction—*people do not just kill other people without some*

> *kind of justification for their consciences*, unless they are simply murderers. The term "savage" and its associated ideology is what the English colonizers used to justify and to rationalize their actions.[84]

Remarkably similar stereotypes reappeared centuries later, during World War II. As historian John W. Dower demonstrates, "It is possible *to obscure what war is about* by denying enemies their humanity and seeing them as beasts, vermin, demons, or devils."[85] For example, President Harry Truman's decision to drop atomic weapons on civilians in Hiroshima and Nagasaki was largely based on his own assumptions that the Japanese people were "savages, ruthless, merciless, and fanatic."[86]

Dehumanizing Stereotypes Specific to American Indians

These same terms—savage, ruthless, merciless, fanatic—had been used for centuries to describe American Indians, even just prior to the outbreak of World War II. For example, even as late as 1939, John Ford's *Stagecoach* and *Drums along the Mohawk* depicted Apaches and Mohawks in these very terms. Dehumanizing depictions were employed in literally hundreds of traditional or "canonical" westerns depicting Native Americans as plagues of vermin.[87]

These can easily be explained in terms of colonizer discourse. Bhabha posits that one of the primary objectives of dehumanizing stereotypes is "to construe the colonized as a population of degenerate types . . . in order to justify conquest."[88] Bhabha's words echo a comment made two decades earlier by Roy Harvey Pearce, who said that Euro-Americans specifically created the image of the savage American Indian in order to justify policies of extermination.[89] American Indian movement leader Russell Means explains that the trick of dehumanizing somehow "makes it all right to kill and otherwise destroy other people. . . . One of the Christian commandments says, 'Thou shalt not kill,' at least not humans, so *the trick is to mentally convert the victims into nonhumans*."[90] In 2004, former Special Forces sergeant Stan Goff describes how Vietnamese civilians too became targets of dehumanization: "We [U.S. forces] had to dehumanize our victims before we did the things we did. . . . We knew deep down what we were doing was wrong, so they became 'dinks' or 'gooks,' just like Iraqis are [now ritually disparaged as] 'ragheads.'"[91]

Colonial Discourse

The ideological smokescreens of the invaders do not always mask the fact that their victims usually have something the colonizers want. Like Bhabha, Pamela Wilson suggests the best way to understand the purposes of negative stereotypes, especially those targeted at American Indians, is to situate them in a context of "colonial discourse."[92] Churchill too insists that the dehumanization of Indians found throughout American literature, film, and television should be seen as part of "an imperial process." In the case of the Americas, Europeans and Euro-Americans wanted the natives' land and its resources.[93]

Benevolent Racism

Stereotypes are not always negative, yet even the "nice" ones are often patronizing and paternalistic. Members of supposedly inferior groups can be *likable*, especially if they are depicted as hapless or subservient. This is the case with Harriett Beecher Stowe's character Uncle Tom. Fiedler notes that Uncle Tom is more or less a variation of Pocahontas.[94] Then there is Uncle Tom's female counterpart, Aunt Jemima, who graces the cover on boxes of pancake mix.[95]

Benevolent racism occurs when the subject group is deemed in need of protection or liberation but is helpless to accomplish this. Members of this group are portrayed as victims of a third, extremely aggressive or immoral group, from which they must be rescued. Sometimes they must be rescued from each other. Such "protection" often takes the form of strong-arm tactics, military occupation, and even genocidal policies against the group being rescued. In the United States, the idea of euphemizing genocidal policies as protection appears to have originated during the Andrew Jackson administration (1829–1837) in the ethnic-cleansing project called Indian Removal. Political scientist Michael Rogin asks, "How to reconcile the elimination of the Indians with the liberal [Euro-American] self-image?" Indeed, this would be a question that preoccupied the statesmen of the period.[96] President Jackson's secretary of war, Lewis Cass, came up with a brilliant way out of this dilemma by formulating the "paternal metaphor," asserting that instead of negotiating, American Indians would fare better had the United States from the start made them its "pupils."[97] The paternal metaphor was a strategy for avoiding cognitive dissonance and sidestepping guilt by turning the truth on its head. This was nothing new, however: the Spanish had invented a similar ideology in the mid-1500s. Frank Tucker writes, "The end result of this philosophy seems to be a mandate for the 'higher' races to conquer 'lower' forms . . . so that those lesser folk can enjoy the superior way of life of the more civilized race."[98]

This was the approach the United States adopted in 1898 when it set out to liberate the Philippines from Spain, only to replace Spain as imperial overlord. Britain's imperial poet Rudyard Kipling, who was living in the United States at the time, urged President William McKinley to "take up the white man's burden." Kipling's ode painted Filipinos as children, dependent on their colonial parents:

> Take up the white man's burden
> Send forth the best ye breed
> Go bind your sons to exile
> To serve your captives' need
> To wait in heavy harness
> On fluttered folk and wild
> Your new-caught, sullen peoples
> Half-devil and half-child[99]

Writing in 1961, Fanon noted the tendency for colonizers to see themselves as parental figures: "Colonialism was not seeking to be perceived as a sweet, kind-hearted mother who protects the child from a hostile environment but rather a mother who constantly prevents her basically perverse child from committing suicide. . . . The colonial mother is protecting the child from itself."[100] Chomsky explained paternalism in plainer language: "When you've got your boot on somebody's neck, you can't just say, 'I'm doing this because I'm a brute.' You have to say, 'I'm doing it because they deserve it. It's for their own good.' . . . They're naughty children who have to be disciplined."[101] The Spanish conquistadores, had, centuries earlier, taken this very idea—"it's for their own good"—to its logical conclusion: killing Native Americans rather than allowing them to continue their "sinful" lives was construed as an act of mercy.[102]

The Good Indian

From the Puritan captivity narratives to Hollywood westerns, American Indians were usually depicted as irrationally hostile, Bloodthirsty Savages. There were cycles of sympathetic, "pro-Indian" portrayals, but until about 1948 the preponderant image was the demonic Bloodthirsty Savage.[103] After the genocides of World War II, however, Hollywood stereotypes depicting the conquest, extermination, and forcible relocation of American Indians by U.S. forces, as seen in countless westerns, became a severe liability on the international stage.[104] The United States, as leader of the Free World, could not afford to be seen as an agent of repression toward nonwhite peoples, especially during a period when many former European colonies were emerging as nations. Hollywood films celebrating the "winning of the West" and the violent victory over American Indians—films exported all over the world—served as reminders that the United States had participated in and *celebrated* its own genocidal practices.[105] This problem was ameliorated when government agents were sent to Hollywood to discourage studios from making movies that would embarrass or hinder the state department's overseas efforts.[106] This effort began during the Truman administration, which is precisely when the television version of *The Lone Ranger* was first aired (1949) and Delmer Daves's film *Broken Arrow* was released (1950). But in fact Hollywood's and the government's objectives had been intertwined for decades: even before World War II, the film industry's Motion Picture Export Agency, a government-sanctioned cartel, called itself "the Little State Department."[107]

U.S. television often used representations of American Indians for political purposes not in the least connected to the interests of Indians themselves. In the 1950s, American Indians were used as "stand-ins" for minorities in general, sometimes for African Americans specifically. This was primarily because, during the televised civil rights clashes of the 1950s, black characters began disappearing from network television.[108] This was not necessarily due to advertisers' racism but more likely to their aversion to any type of public controversy, and the issue of civil rights was a hot potato no one wanted to touch.[109] According to *Variety* columnist Dave Kaufman, race was "too strong" an issue for commercial television.[110] When a television writer or producer wanted to address the disturbing issue of segregation or racism, it had to be done allegorically. Because battles with American Indians were far in the past, they were considered a safe minority and could be used as stand-ins for blacks in order to insinuate pluralist-liberal critiques of segregation while avoiding alarming advertisers.[111]

In 1960 the "model-minority" stereotype emerged that would recur at points when race relations between whites and African Americans became tense. Asians and Asian Americans were the groups most often used as the model minority, but American Indians would be used for this purpose as well. Also during the 1960s, American Indians became stand-ins for disaffected white youth. Hippies and counterculture types, who identified with American Indians—to a point—fashioned faux-Indian clothing and shoulder-length hair into their own critique of modernist, industrial society.[112] Ken Kesey used Chief Bromden for this purpose in his novel *One Flew over the Cuckoo's Nest* (1962; adapted to film in 1975 by Milos Forman). After the 1968 My Lai massacre, director Arthur Penn used American Indian characters as metaphorical substitutes for Vietnamese peasants in *Little Big Man* (1970; adapted from a 1964 novel by Thomas Berger), drawing a loose comparison of U.S. policies during the winning of the West to scorched-earth policies in Vietnam.[113] Kevin Costner's 1990 film *Dances with Wolves* made a similar connection as did James Cameron's 2009 film *Avatar*. However well-intentioned the motives behind such films, the fact remained that American Indians were being used as political puppets to illuminate critiques of U.S. foreign policy, just as centuries earlier the French philosophes had appropriated the Noble Savage stereotype as a polemical device for critiquing their own society.

Other "Positive" Stereotypes of American Indians

In 1970, American Indian scholar and cultural critic Vine Deloria Jr. suggested that Indians had more rational lifestyles than Euro-Americans, whose rampant technologies, consumerism, and scorched-earth military policies tended to destroy the environment.[114] Indians came to be used as "poster children" for environmental causes: modern whites "embraced the image of the Native American as the ultimate environmentalists."[115] This stereotype does have some basis in American Indian religion and cosmology. The most prominent example was a well-known public service announcement (PSA) featuring Sicilian-American actor Iron Eyes Cody (real name Esperi DiCorti) shedding a tear over the mess Anglo-American civilization had wreaked on the once-beautiful landscape. This famous PSA, which first ran on Earth Day in 1971, was part of the Advertising Council's Keep America Beautiful campaign.[116]

An American Indian in buckskins sallies upriver in a canoe (figure 0.2), only to land in a pile of litter; he then confronts a massive traffic jam and a skyline of smokestacks and smog. The camera finally moves in for an extreme close-up of a tear falling down his cheek (figure 0.3). While it is true

Figure 0.2. Opening sequence from Keep America Beautiful public service announcement (1971).

Figure 0.3. Iron Eyes Cody as the "crying Indian" (same public service announcement).

that most Native American groups do attach religious significance to the land, the problem with this trope is that, while it presents the Indian in a sympathetic, romanticized light, it locks him into the past, making him a symbol of a forgotten era—and a forgotten people.[117]

As noted earlier, negative and positive stereotypes operate simultaneously. The Puritan leaders of early Massachusetts were quick to paint the "heathen" natives as demonic and/or satanic—yet they also regarded the natives who helped them as divine messengers sent by Providence.[118] One reason both the Bloodthirsty Savage and Noble Savage stereotypes have operated together simultaneously is because U.S. ideologues wanted to position (or synthesize) the new nation somewhere between what they saw as an overcivilized, effete Europe and the "savage" Native America; they wanted to be a bit like Indians themselves.

Whether one advocated extermination or pity, American Indians were in any case doomed to extinction.[119] This doomed-race trope, probably derived from Cooper's *Last of the Mohicans* (1826), was repeated in many films (as in, for example, D. W. Griffith's adaptation of Helen Hunt Jackson's *Ramona*) as well as on television more than a century later. For example, in the 1988 series *Guns of Paradise*, the chief and last surviving member of the Paiute nation (reminiscent of Cooper's Chingachgook), John Taylor (Dehl Berti), insists that the extinction of his people is inevitable and that any struggle against Euro-American domination is useless.[120] Even Russell Means invokes this trope in a 1992 film version of *Last of the Mohicans*. This doomed-race trope, like a Barthesian compromise, allows Euro-Americans to feel both contrition and self-satisfaction at the same time.

PROLIFERATION OF WESTERNS ON U.S. TELEVISION DURING THE COLD WAR ERA

Westerns had been popular fare since the dime novel made its debut in the 1860s, and western films had been the most prevalent genre in U.S. film since the dawn of the industry.[121] Indeed, the very first films made in the United States could be deemed westerns. As Anglo-America's foundation myth, the western often celebrated war and the seizing of the land from its aboriginal inhabitants, at least until World War II, when the Hollywood western was rehabilitated. Westerns were, in any case, celebrations of masculinity; indeed, Lutz Koepnick describes the western as "phallic rhetoric."[122] With their martial settings, westerns lent themselves to pro-war propaganda—although there have been some notable westerns that subverted the genre, evincing *anti*war sentiments—so it is not surprising they proliferated in the period before and during World War II. Tom Pollard writes, "Westerns merely substituted western landscapes and had Indians stand in for Nazis or Huns or Japs of traditional combat films. Both the western and combat genres, of course, are male-dominated, both stress extremely violent action, and both feature organized groups of heroes taking on a militarized enemy."[123]

After the war, westerns faded in prominence for a time (*The Lone Ranger* radio series notwithstanding), but the Cold War brought a major resurgence: westerns proved ideal vehicles for Cold War ideology as well as ideas about leadership and masculinity and the relationship between the two.[124] In 1952, the year Dwight D. Eisenhower was elected president, CBS Radio trotted out *Gunsmoke*, whose protagonist was a fictional U.S. marshal from Dodge City, Kansas, Eisenhower's home state.[125] Eisenhower let it be known that he was a fan of westerns: In a 1953 speech he intoned,

> I was raised in a little town of which most of you have never heard. But in the West it is a famous place. It is called Abilene, Kansas. We had as our marshal for a long time a man named Wild Bill Hickok. If you don't know anything about him, read your westerns more.[126]

During Eisenhower's two terms (1953–1960), the quantity of television westerns jumped from 11 in 1953 to 50 in 1959 (see figure 0.4). By 1958, there were at least two westerns appearing every night of the week.[127]

Because nationwide network broadcasting in the United States did not materialize until 1951, television depictions of American Indians followed a different path than had evolved in film.[128] Television's portrayals, governed according to postwar political necessities, avoided film's negative stereotypes in favor of the Good Indian exemplified by Tonto. Bad Indians were rarely seen on network television (although they could be seen in reruns of B westerns on local stations). The Good Indian would become a staple in network-television programs, especially in those featuring an Indian in a prominent or starring role.

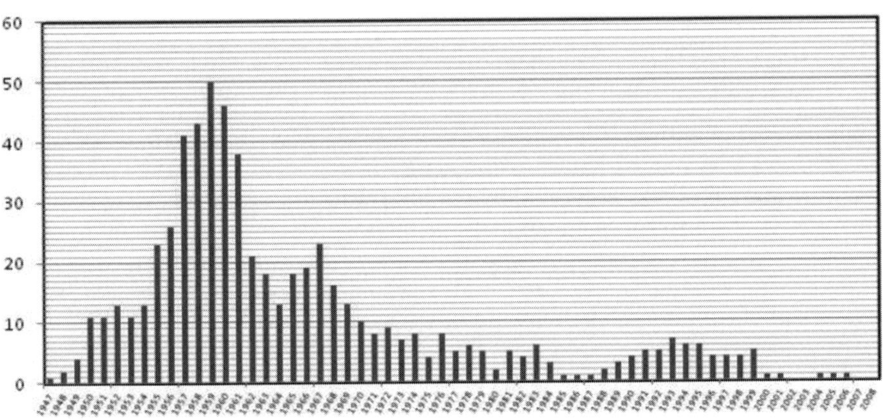

Figure 0.4. U.S. television westerns, 1947–2007. The peak occurred during Eisenhower's second term. Author's research.

INSTITUTIONAL ISSUES

During the Cold War, the Big Three network oligopoly collaborated with the government to disseminate nationalist ideals and promote national unity; programs featured "American heroes who fought against national threats and reinforced positive images of the US government and its people."[129] The networks, in short, behaved as "adjuncts of government."[130] As Althusser predicts, the major media outlets behaved as Ideological State Apparatuses, whether state owned or private, not because of some dark conspiracy or government intervention (which did, however, occur[131]) but because they had intersecting interests, which in the late 1940s and early 1950s included building up the national-security state and expanding defense budgets.[132] The networks had a stake in the defense budgets since the parent companies of all three networks at one time or another—General Electric (GE), Radio Corporation of America, and Westinghouse Electric—had been defense contractors; not only were they in bed with the military but they were dependent on revenues from government contracts.[133] NBC's majority owner until 2011, GE, is still a top defense contractor.[134] Although CBS and ABC were ostensibly independent, CBS was managed primarily by Frank Stanton, who doubled as CEO of RAND Corporation, a defense contractor with close ties to the military. Thus what is commonly referred to as the "military-industrial complex"—a term used by President Eisenhower in his 1961 farewell speech—should more appropriately be called the "military-industrial-*media*-complex" or "military-industrial-media-entertainment (MIME) complex."[135]

U.S. television is an advertising medium, so there is always the issue of how television images attract viewers and help sell the products lurking in the background. The method it uses to do so is largely based on classical conditioning: it provides viewers with a "positive affect" or a "warm, fuzzy" feeling, which advertisers hope viewers will "attach" to the products and services featured in advertisements.[136] This was based on research conducted on radio broadcasting in the 1920s. AT&T, one of the pioneers of the U.S. radio industry, conducted studies of advertising and found that simply *attaching the name of a product to a pleasant experience* would associate it with the good feelings generated by the program.[137]

One of the stereotype's appeals is to position its user, in an implied structural comparison, as naturally *superior to* the "object" being depicted: the user places him- or herself *above* the object in the "marked category," giving the subject a feeling of superiority. Propagandistic images, writes journalist Chris Hedges, "appeal to our deepest-held, most primitive prejudices, our classism, sexism, racism—perversions based on fear of complexity or change. Propaganda contains much of what we already yearn to believe."[138] This positions viewers on the "correct side" of the question and provides a good feeling or "positive affect," in advertising-research terms. If this feeling of correctness might be *transferred* (or "attached") to the products in the advertisements, it could help sell them. The use of stereotypes, then, attempts to sell products by positioning the viewer as superior, giving him or her an "ego-boost," which is subliminally perceived as pleasurable and supposedly provides encouragement for the viewer to purchase the products associated with the program. This offers one explanation as to why U.S. television relies heavily on stereotypes.[139]

CONCLUSION

This introduction has described some of the traditional stereotypes used with regard to Native Americans and how they evolved in the United States. It has demonstrated how these were part of an imperial process, which the United States itself tends to deny, and how it must devise elaborate ideological mechanisms to perpetuate this denial.[140] It has also demonstrated that the Hollywood-studio oligopoly, the television-network oligopoly, the major defense contractors, and the government had intertwining interests and ideologies. However, it must be noted that these players do not comprise a collective, monolithic entity: they are more like shifting feudal alliances or what Hall calls "historic formations." Such negotiating and compromising strategies correspond with Hall's assertion that the mass media comprise an arena where battles for meaning are fought and resolved.[141] Media owners, then, did not have *all* the power to create meaning, but they certainly owned and controlled the mechanisms—whether the public accepted what was offered is another matter.

During the Cold War it was necessary to create a vision of a pluralist, multicultural America, one in which minorities and subaltern groups were ostensibly welcomed, and the networks collaborated in this project. It was one thing to watch this multicultural fable on programs like *The Lone Ranger*, *Broken Arrow*, and *Law of the Plainsman*; it was quite another for Americans to actually look around and see what was taking place in their cities. Many if not most wanted so badly to believe that the United States was a fair and equitable society—as their leaders had promised—that they refused to see the inequities in their inner cities; it proved far easier to simply leave for the suburbs, where they could escape the contradictions of the nation's social and economic policies and live in a televisual *simulacrum* where "father knows best." Meanwhile, poor African Americans were left behind in urban ghettoes, while Native Americans remained on out-of-the-way reservations and suffered the worst living conditions of any group.[142]

NOTES

1. Walter Lippmann, *Public Opinion* (New Brunswick, NJ: Transaction Publishers, 1998), 96.

2. Edward Said, *Culture and Imperialism* (New York: Vintage Books, 1993), 52.

3. I use the terms "American Indian" and "Native American" interchangeably. However, American Indians themselves generally prefer the latter term.

4. Robert F. Berkhofer Jr., *The White Man's Indian: Images of the American Indian from Columbus to the Present* (New York: Alfred A. Knopf, 1978), xv, 10.

5. James Fenimore Cooper, *The Leatherstocking Tales*. Volume 1 (New York: Library of America, 1985), 479.

6. "The Angry American Indian: Starting Down the Protest Trail," *Time*, February 9, 1970. http://www.time.com/time/magazine/article/0,9171,876550,00.html

7. Daniel Francis, *The Imaginary Indian: The Image of the Indian in Canadian Culture* (Vancouver, BC: Arsenal Pulp Press, 1992), 107–8.

8. See, for example, Berkhofer, 1978, 76–80, 90–92.

9. Leslie Fiedler, *The Return of the Vanishing American* (New York: Stein & Day, 1968), 70.

10. Quoted in "Pocahontas Legend Hides Unromantic Realities." *Los Angeles Times*, June 24, 1995. http://articles.latimes.com/1995-06-24/local/me-16637_1_historical-pocahontas.

11. Fiedler, 1968, 70.

12. The term "ethnic sidekick" comes from Frederick Zackel, "Robinson Crusoe and the Ethnic Sidekick." *Bright Lights Film Journal* 58, November 2007. http://www.brightlightsfilm.com/30/crusoe1.html.

13. Peter Hulme, *Colonial Encounters: Europe and the Native Caribbean* (London: Methuen, 1986), 46–47, 50, 53, 65. I should also mention Patuxet native Squanto or Tisquantum, who served as a lieutenant for John Smith but was captured by other Englishmen and brought to Spain to be sold into slavery until some Spanish priests intervened. He eventually made his way back to North America via England about 1621 only to find his kinsmen dead of disease. He continued to work as a guide and translator for the English. Disney made a film about his life in 1994, titled *Squanto: A Warrior's Tale*, starring Adam Beach. Squanto's story fits well into the man Friday model this book argues remains the predominant role of American Indians in Euro-American culture.

14. Mary Alice Money, "Broken Arrows: Images of Native Americans in the Popular Western." In Dane Morrison, ed. *American Indian Studies: An Interdisciplinary Approach to Contemporary Issues* (New York: Peter Lang, 1997), 365–66.

15. Prime examples would be the middle-class domestic comedies of the 1950s, especially *Leave It to Beaver*, in which the uncomfortable issue of minorities—and their rights—is dispelled: there simply are no minorities.

16. George Gerbner and Larry Gross, "Living with Television: The Violence Profile." *Journal of Communication* 26(2), June 1976, 182.

17. Cedric C. Clark, "Television and Social Control: Some Observations on the Portrayals of Ethnic Minorities." *Television Quarterly*, Spring 1969, 18–22. Clark also mentions this in a 1973 chapter, "The Concept of Legitimacy in Black Psychology." In Edgar G. Epps, ed. *Race Relations: Current Perspectives* (Cambridge, MA: Winthrop), 332–54.

18. Dyer comes close to this approach in his extensive study of how European and Euro-American art and film tend to deify Euro-Americans themselves, but he does not specifically discuss television depictions. Richard Dyer, *White* (London: Routledge, 1997).

19. Roland Barthes, *Mythologies* (London: Paladin, 1972), 11.

20. *The Winning of the West* was the title of a series of history books penned by Theodore Roosevelt, in which he celebrates the bravery of the Euro-American "race" in seizing the continent. Theodore Roosevelt, *The Winning of the West*. Volume 1, *Spread of the English-Speaking Peoples* (New York: G. P. Putnam's Sons, 1910).

21. I offer apologies to American Indian scholars who might feel I have given this aspect short shrift.

22. For a detailed discussion on the subaltern, see Gayatri Chakravorty Spivak, "Can the Subaltern Speak?" In Cary Nelson and Lawrence Grossberg, eds. *Marxism and the Interpretation of Culture* (Champaign: University of Illinois Press, 1988): 271–315.

23. George Gerbner, *Women and Minorities on Television: A Study in Casting and Fate.* Screen Actors Guild report, 1993. http://www.asc.upenn.edu/gerbner/Asset.aspx?assetID=333; also see Diana George and Susan Sanders, "Reconstructing Tonto: Cultural Formations and American Indians in 1990s Television Fiction." *Cultural Studies* 9(3), 1995, 427–52.

24. Ella Shohat, "Gender and the Culture of Empire." In Hamid Naficy and Teshone Gabriel, eds. *Otherness and the Media: An Ethnography of the Imagined and the Imaged* (Langhorne, PA: Harwood Academic Publishers, 1993), 75.

25. *The Lone Ranger.* Episode 1: "Enter the Lone Ranger" (Apex Film Corp./ABC-TV, September 15, 1949).

26. Herman Gray, "The Politics of Representation in Network Television." In Horace Newcomb, ed. *Television: The Critical View*, 6th ed. (Oxford: Oxford University Press, 2000), 294–95; also see Naficy and Gabriel, ix.

27. Gray in Newcomb, 2000, 294.

28. I suspect that this development in the evolution of the American Indian archetype had been influenced by the character Superman, who had been popular in comic books since 1938 and was appearing on television at this time as well. *Adventures of Superman* (Motion Pictures for Television, 1952–1958).

29. Will Wright, *Six Guns and Society* (Berkeley: University of California Press, 1977), 58, 82. Also see Lee Clark Mitchell, *Westerns: Making the Man in Fiction and Film* (Chicago: University of Chicago Press, 1996), 159.

30. Dyer, 1997, 146.

31. Gray in Newcomb, 2000, 294–95.

32. Daniel Chirot and Clark McCauley, *Why Not Kill Them All? The Logic and Prevention of Mass Political Murder* (Princeton, NJ: Princeton University Press, 2010), 54.

33. I would like to have employed psychoanalytical methods as well, but these are far too complex and perhaps too controversial to include here.

34. Alex McNeil, *Total Television: The Comprehensive Guide to Programming from 1948 to the Present*, 4th ed. (New York: Penguin Books, 1996); Tim Brooks and Earl Marsh, *The Complete Directory to Prime Time Network and Cable Television Shows* (New York: Random House, 2003). The TV Acres website, compiled by retired California State University reference librarian, Jerome A. Holst, has not been vetted by an authority, but I have found it to be thorough and reliable.

35. Stuart Hall, "Notes on Deconstructing the Popular." In Raphael Samuel, ed. *People's History and Socialist Theory* (London: Routledge, 1981), 232–33.

36. Jacques Aumont, Alain Bergala, Michel Marie, and Mark Vernet, eds., *Aesthetics of Film* (Austin: University of Texas Press, 1992), 117–18.

37. Said, 1993, 80.

38. Barthes, 1972.

39. John Storey, ed., *Cultural Theory and Popular Culture: A Reader* (Athens: University of Georgia Press, 1998), xi–xv.

40. Antonio Gramsci, *Selections from the Prison Notebooks* (New York: International Publishers, 1971; written 1929).

41. Todd Gitlin, "Prime Time Ideology: The Hegemonic Process in Television Entertainment." In Horace Newcomb, ed. *Television: The Critical View* (Oxford: Oxford University Press, 2007), 507–32.

42. Edward S. Herman and Noam Chomsky, *Manufacturing Consent: The Political Economy of the Mass Media* (New York: Pantheon Books, 1988). Edward L. Bernays, *The Engineering of Consent* (Norman: University of Oklahoma Press, 1969). The term "hegemony" in a cultural sense emanates from Antonio Gramsci. Thomas R. Bates, "Gramsci and the Theory of Hegemony." *Journal of the History of Ideas* 36(2), April 1975, 351–66. It is basically a modification of Marx's idea that the values of the ruling class become the values of the society as a whole. It was revived in the 1980s by media sociologist Todd Gitlin.

43. Theodor Adorno, *The Culture Industry* (London: Routledge, 2006). Herbert Marcuse, *One-Dimensional Man* (Boston: Beacon Press, 1964). Rolf Wiggershaus, *The Frankfurt School: Its History, Theories, and Political Significance* (Boston: MIT Press, 1995).

44. Louis Althusser, *Lenin and Philosophy* (London: NLB, 1971), 150.

45. Robert McChesney, *Rich Media, Poor Democracy: Communications Politics in Dubious Times* (New York: The New Press, 1999); also see McChesney, *Ruthless Criticism: New Perspectives in US Communication History* (Minneapolis: University of Minnesota Press, 1993).

46. Gerbner and Gross, 1976; also see Aumont et al., 1992.

47. Storey, 1998, xi–xv.

48. Stuart Hall, "Notes on Deconstructing the Popular," in Raphael Samuel (ed.). *People's History and Socialist Theory* (London: Routledge, 1981), 232.

49. Ibid., emphasis added.

50. Storey, 1998, xiv.

51. Kalle Lasn, *Culture Jam: The Uncooling of America* (New York: William Morrow, 1999), xiii.

52. Lippmann, 1998, 79–103.

53. Herbert I. Schiller, *The Mind Managers: How the Master Puppeteers of Politics, Advertising, and Mass Communication Pull the Strings of Public Opinion* (Boston: Beacon Press, 1973), 106–7, 173, 188; Christopher Simpson, *Science of Coercion: Communication Research and Psychological Warfare, 1945–1960* (Oxford: Oxford University Press, 1994), 4, 13–15, 19–20.

54. Glen Ford, "Corporate Media's Crimes against Libya—and Humanity." *Black Agenda Report*, June 8, 2011. http://www.trinicenter.com/modules.php?name=AvantGo&op=ReadStory&sid=2332.

55. Simpson, 1994, 15–30, 42–51.

56. Richard Dyer, "The Role of Stereotypes." In Paul Marris and Sue Thornham, eds. *Media Studies: A Reader* (New York: New York University Press, 2000), 245.

57. Ward Churchill, *Fantasies of the Master Race: Literature, Cinema, and the Colonization of American Indians* (San Francisco: City Lights Books, 1998), 9; also see Homi K. Bhabha, "The "Other" Question: Difference, Discrimination, and the Discourse of Colonialism." *Screen* 24(6), 1983, 18–22.

58. Montesquieu (Charles de Secondat), *The Spirit of the Laws* (New York: Free Press, 1970), 239.

59. John Sabini, *Social Psychology* (New York: W. W. Norton, 1992), 626–67, 634–35; Leon Festinger, *A Theory of Cognitive Dissonance* (Stanford, CA: Stanford University Press, 1957).

60. Robin D. G. Kelly, "A Poetics of Anticolonialism." *Monthly Review* 5(6), November 1999. http://www.monthlyreview.org/1199kell.htm.

61. Frantz Fanon, *The Wretched of the Earth* (New York: Grove Press, 2004); also see *Black Skin, White Mask* (New York: Grove Press, 2008).

62. Barthes, 1972, 115; emphasis added.

63. Ibid., 129.

64. Ibid.

65. W. E. B. DuBois, *Darkwater: Voices from within the Veil. The Oxford W.E.B. DuBois Reader*. Edited by E. J. Sundquist (Oxford: Oxford University Press, 1996), 59–60.

66. James Baldwin, *The Fire Next Time* (New York: Dial Publishing, 1963), 599.

67. Fiedler, 1968, 44, 50.

68. Richard Drinnon, *Facing West: The Metaphysics of Indian-Hating and Empire-Building* (Minneapolis: University of Minnesota Press, 1980).

69. Michael Paul Rogin, "Liberal Society and the Indian Question." *Politics and Society*, May 1971, 269–312. Reprinted in *Ronald Reagan, the Movie and Other Episodes in Political Demonology* (Berkeley: University of California Press, 1987), 134–68.

70. Churchill, 1998.

71. Dyer, 1997.

72. Reginald Horsman, *Race and Imperialism* (Cambridge, MA: Harvard University Press, 1981), 17.

73. Lippmann, 1998, 96.

74. *Birth* was based on Thomas F. Dixon's 1905 novel *The Clansman*.

75. Berkhofer, 1978, 76–77.

76. Bruce E. Johansen, *Forgotten Founders: How the American Indian Helped Shape Democracy* (Cambridge, MA: Harvard Commons Press, 1987).

77. Berkhofer, 1978, 77–78.

78. Jeff McArn, "Reflections on Political, Religious Uses of the Term 'Evil.'" *Observer-Dispatch* (Utica, NY), June 13, 2004. http://collegenews.org/editorials/1980/reflections-on-political-religious-uses-of-the-term-evil.html.

79. C. S. Cowles, Stanley N. Gundry, Daniel L. Gard, Eugene H. Merrill, and Tremper Longman, *Show Them No Mercy: Four Views of Canaanite Genocide* (Grand Rapids, MI: Zondervan, 2003), 97–98.

80. Quoted in Tzvetan Todorov, *The Conquest of America: The Question of the Other* (New York: Harper and Row, 1984), 153.

81. Ibid., 150.

82. David Stannard, *American Holocaust: The Conquest of the New World* (Oxford: Oxford University Press,1992), 95.

83. Bhabha, 1983, 18–36.

84. Linwood Custalow and Angela L. Daniel, *The True Story of Pocahontas* (Golden, CO: Fulcrum Publishing, 2007), 36; emphasis added.

85. John W. Dower, *War without Mercy: Race and Power in the Pacific War* (New York: Pantheon, 1986), 294; emphasis added.

86. Harry S. Truman, diary entry July 25, 1945; quoted in Eduard Mark, "'Today Has Been a Historical One': Harry S Truman's Diary of the Potsdam Conference." *Diplomatic History* 4(3), Spring 1980, 317–26.

87. Although, as Angela Aleiss demonstrates, there have been cycles of "pro-Indian" films throughout U.S. film history. Angela Aleiss, *Making the White Man's Indian: Native Americans and Hollywood Movies* (New York: Praeger Publishers, 2005).

88. Bhabha, 1983, 18–36.

89. Roy Harvey Pearce, *Savagism and Civilization: A Study of the Indian and the American Mind* (Berkeley: University of California Press, 1964), 105–27.

90. Russell Means, "Marxism Is a European Tradition." Speech given at Black Hills International Survival Gathering, Rapid City, SD, July 1980. Reprinted in "Fighting Words on the Future of the Earth." *Mother Jones*, December 1980, 26–27; emphasis added.

91. Stan Goff, "An Open Letter to GIs: Hold on to Your Humanity." *International Socialist Review*, January 2004. http://isreview.org/issues/33/humanity.shtml.

92. Churchill, 1998, 56.

93. Ibid., 9.

94. Fiedler, 1968, 70.

95. The model for Aunt Jemima was Nancy Green, who was born into slavery in 1834 and died in 1923. "Nancy Green, the Original Aunt Jemima." African American Registry. http://www.aaregistry.com/african_ american_history/1287/Nancy_Green_the_original_Aunt_Jemima.

96. Rogin, 1987, 136.

97. Frank B. Woodford, *Lewis Cass: The Last Jeffersonian* (New York: Octagon Books, 1973), 143.

98. Frank H. Tucker, *The White Conscience: A Hard-Hitting, Documented Survey of 500 Years of Western Domination* (New York: Frederick Ungar, 1968), 35.

99. Rudyard Kipling, "(Take Up) The White Man's Burden." *McClure's* 12, February 12, 1899. Reprinted at World Civilizations Reader. http://www.wsu.edu:8080/~wldciv/world_civ_reader/ world_civ_reader_2/ kipling.html.

100. Fanon, 2004, 149.

101. Quoted in David Barsamian, "Telling the Truth about Imperialism." *International Socialist Review*, November–December 2003, 18.

102. Todorov, 1996, 218; also see Anne DeLong, "Disrupting the Discourse of Conquest: The Suppression of Sepúlveda." The Literature of Justification. Accessed August 1, 2011. Lehigh University Digital Library. http://digital.lib.lehigh.edu/trial/justification/newspain/essay/.

103. Aleiss, 2005; also see Ward Churchill, Mary Anne Hill, and Norbert S. Hill Jr., "Examination of Stereotyping: An Analytical Survey of Twentieth Century Indian Entertainers." In Gretchen M. Bataille and Charles L. P. Silet, eds. *The Pretend Indians: Images of Native Americans in the Movies* (Ames: Iowa State University Press, 1980), 35–48.

104. U.S. Department of State Bulletin. May 20, 1963, 783. Cited in Tucker, 1968, 334, fn 11.

105. Thomas Borstelmann, *The Cold War and the Color Line: American Race Relations in the Global Arena* (Cambridge, MA: Harvard University Press, 2001). Mary L. Dudziak, *Cold War Civil Rights: Race and the Image of American Democracy* (Princeton, NJ: Princeton University Press, 2000); also see U.S. Department of State Bulletin in Tucker, 1968, 11 .

106. David Eldridge, "'Dear Owen': The CIA, Luigi Luraschi and Hollywood, 1953." *Historical Journal of Film, Radio, and Television* 20(2), June 2000, 151–56.

107. Toby Miller, "Hollywood and the World." In John Hill and Pamela C. Gibson, eds. *The Oxford Guide to Film Studies* (Oxford: Oxford University Press, 1998), 372.

108. Melvin P. Ely, *The Adventures of Amos 'n' Andy: A Social History of an American Phenomenon* (New York: Free Press, 1991), 228.

109. Robin Means-Coleman, *African American Viewers and the Black Situation Comedy: Situating Racial Humour* (New York: Garland Books, 2000), 82.

110. Dave Kaufman, quoted in Pepe Arciga, *Los Angeles Mirror*, August 26, 1957. Reprinted in *Los Angeles Times*, August 17, 2007. http://latimesblogs.latimes.com/thedailymirror/2007/08/pepe-arciga-1.html#more.

111. Philip French, *Westerns* (New York: Viking Press, 1973), 81, 95. For example, award-winning writer Rod Serling, hoping to make a point about racial prejudice, was forced to rewrite an African American character as a Jew: the script for Serling's "Noon on Doomsday" for the *US Steel Hour* (CBS, January 1956) turned fourteen-year-old lynching victim Emmett Till into an elderly Jewish pawnbroker. Jon Kraszewski, "Adapting Scripts in the 1950s: The Economic and Political Incentives for Television Anthology Writers." *Journal of Film and Video* 58(3), Fall 2006, 3–22. Also see Mary Ann Watson, *The Expanding Vista: American Television in the Kennedy Years* (Durham, NC: Duke University Press, 1994), 57. However this phenomenon had been noted by contemporaneous observers even as it was taking place; see Donald Shea Teeple, "TV Westerns Tell a Story." *American Mercury,* April 1958, 115. The term "stand-in" in this particular context seems to have originated with Thomas Cripps, *Making Movies Black: The Hollywood Message Movie from World War II to the Civil Rights Era* (Oxford: Oxford University Press, 1993), 281.

112. Philip Deloria, *Playing Indian* (New Haven, CT: Yale University Press, 1998), 161; also see "Angry American Indian," 1970; Rogin, 1987, 322, fn 4; Todorov, 1984, 251.

113. Ronald Bergan, "In Praise of Arthur Penn." *Guardian* (United Kingdom), February 14, 2007. http://www.guardian.co.uk/film/filmblog/2007/feb/14/arthurpenn. Literary historian Richard Drinnon, in his historical study of U.S. literature and film, later drew much the same conclusion: the My Lai massacre was, for all practical purposes, where the trajectory of U.S. foreign policy had been headed since Puritan days. Drinnon, 1980, 452–60.

114. Vine Deloria Jr., *We Talk You Listen: New Tribes, New Turf* (Lincoln, NE: Bison Books, 2007), 186.

115. David Rich Lewis, "Still Native: The Significance of Native Americans in the History of the Twentieth Century American West." *Western Historical Quarterly* 2(24), May 1993, 221.

116. "KAB Public-Service Announcements." Keep America Beautiful. n.d. https://secure2.convio.net/kab/site/SPageServer?pagename=kabmedia_psaandJServSessionIdr006=iqn2rv6652.app6b.

117. Lewis, 1993, 221.

118. Bataille and Silet, eds., *The Pretend Indians: Images of Native Americans in the Movies* , xxi; Donald L. Kaufman. "The Indian as Media Hand-Me-Down." *Colorado Quarterly* 23, Spring 1975, 489–504; Drinnon 1980, 18, 38, 51–52; Berkhofer, 1978, 81–83. Ralph Friar

and Natasha Friar, *The Only Good Indian: The Hollywood Gospel* (New York: Drama Book Specialists, 1972).

119. Berkhofer, 1978, 90–93.

120. George and Sanders, 1995, 435.

121. Aleiss, 2005.

122. Lutz P. Koepnick, "Siegfried Rides Again: Westerns, Technology, and the Third Reich." *Cultural Studies* 11(3), 1997, 418.

123. Tom Pollard, "The Hollywood War Machine." In Carl Boggs, ed. *Masters of War: Militarism and Blowback in the Era of American Empire* (London: Routledge, 2003), 328.

124. Stanley Corkin, *Cowboys as Cold Warriors* (Philadelphia: Temple University Press, 2004). J. Fred MacDonald, *Who Shot the Sheriff? The Rise and Fall of the Television Western* (Westport, CT: Praeger, 1987); J. Fred MacDonald, *Television and the Red Menace: The Video Road to Vietnam* (Westport, CT: Praeger, 1985), 134–42.

125. CBS chairman William Paley had wanted to develop a "Philip Marlowe in the old West," that is, a western-styled imitation of *The Adventures of Philip Marlowe*, a popular CBS detective series, since the late 1940s (Shreve n.d.). Ivan G. Shreve, "Premier Collections: The Adventures of Philip Marlowe, Volume One" (liner notes). *Radio Archives.Org*. http://www.radioarchives.org/sets/PC29.htm. Paley's idea was to synthesize the western and crime genres, thereby creating the "adult western." *Gunsmoke*'s setting in Eisenhower's home state may have been more than coincidence, since according to Paley's autobiography, he and Eisenhower were close friends and golf partners. During World War II, Paley had served on Eisenhower's staff; he later became one of Eisenhower's financial supporters and served as his unofficial media advisor. William S. Paley, *As It Happened: A Memoir* (New York: Doubleday, 1979), 92, 311.

126. Ironically, even though Eisenhower and Adolph Hitler fought on opposite sides during World War II, they both were fans of westerns. Koepnick, 1997, 419. Hitler had been an ardent devotee of German author Karl May. Klaus Mann, "Cowboy Mentor of the Fuhrer." *Living Age* 359, November 1940, 217–18.

127. Brooks and Marsh, 2003, 1365.

128. Norman A. Scotch, "The Vanishing Villains of Television." *Phylon* 21(1), Fall–Winter 1960, 58–62.

129. Tricia Jenkins, "Nationalism and Gender: The 1970s, *The Six Million Dollar Man*, and *The Bionic Woman*." *Journal of Popular Culture* 44(1), February 2011, 95.

130. The term "adjuncts of government," which clearly has an Althuserrian ring, was coined by Sen. William J. Fulbright, referring to the U.S. press and was later used by Noam Chomsky in his book *Necessary Illusions*. William J. Fulbright, Remarks given on the Senate floor. U.S. Congress. Senate. Committee on Foreign Relations. August 1966. Reprinted at Center for Research Libraries. http://images.crl.edu/178.pdf. Noam Chomsky, *Necessary Illusions: Thought Control in Democratic Societies* (Boston: South End Press, 1989), 75–104.

131. Carl Bernstein, "The CIA and the Media." *Rolling Stone*, October 20, 1977: 57–61.

132. Andrew Bacevich, "The Tyranny of Defense Inc." The Atlantic, January/February 2011. http://www.theatlantic.com/magazine/archive/2011/01/the-tyranny-of-defense-inc/8342; John Bellamy Foster, Hannah Holleman, and Robert W. McChesney, "The US Imperial Triangle and Military Spending." Monthly Review 60(5), October 2008. http://monthlyreview.org/081001foster-holleman-mcchesney.php.

133. In 1919, the U.S. Navy encouraged the formation of a consortium of the three top electrical manufacturers, General Electric (formerly Edison General Electric), Westinghouse Electric, and American Telephone and Telegraph (AT&T), to monopolize broadcast technologies in the name of national defense. These three companies along with United Fruit Co., with government backing, formed Radio Corp. of America, of which GE was the largest shareholder with 35 percent. AT&T left RCA voluntarily in 1925, and GE was forced to sell its interest in 1930 due to an antitrust ruling. However, when RCA formed National Broadcasting Co. in 1926, GE and Westinghouse came in as NBC's co-owners. NBC took over the radio networks that GE, Westinghouse, and AT&T had developed, which were renamed NBC Red and NBC Blue. NBC was forced to sell its Blue network in 1943, whereupon it was renamed the American Broadcasting Company (ABC). In 1995, the Columbia Broadcasting System (CBS),

which had hitherto been independent, was taken over by Westinghouse. William Sloan and David Startt, *The Media in America: A History* (Northport, AL: Vision Press, 1999), 341–44. In 1999 CBS-Westinghouse merged with media conglomerate Viacom.

134. Eloise Lee and Robert Johnson, "The 25 Biggest Defense Companies in America." *Business Insider*, March 13, 2012. http://www.businessinsider.com/top-25-us-defense-companies-2012-2?op=1.

135. Marianna Torgovnick, *The War Complex: World War II in Our Time* (Chicago: University of Chicago Press, 2005), 16. Norman Solomon, "The Military-Industrial-Media Complex: Why War Is Covered from the Warriors' Perspective." Fairness and Accuracy in Reporting. August 2005. http://www.fair.org/index.php?page=2627. James der Derian, *Virtuous War: Mapping the Military-Industrial-Media-Entertainment Complex* (New York: Routledge, 2009), xxxv–xxvi, 243.

136. Tony Yeshin, *Advertising* (Florence, KY: Cengage Learning Business Press, 2005), 3, 378. However, the reverse might also work: a violent scene might agitate viewers, and the *relief* the advertisement provides from this could stimulate a positive affect.

137. Randy D. Jacobs, "Advertising Agencies and the Adoption of Radio: A Diffusion of Innovations Perspective." Paper presented at Conference on Historical Analyses and Research in Marketing, New York, May 22, 2011.

138. Chris Hedges, *American Fascists: The Christian Right and the War on America* (New York: Free Press, 2006).

139. Gerbner, 1993.

140. Conservative commentators Charles Krauthammer, Max Boot, William Kristol, and Canadian Michael Ignatieff are exceptions. They suggest conservatives celebrate the idea of an American Empire in the same manner the British once celebrated theirs. Emily Eakin, "All Roads Lead to D.C." *New York Times*, Week in Review, March 31, 2002. http://www.nytimes.com/learning/teachers/featured_articles/20020401monday.html; also see Maximilian C. Forte, ed. *The New Imperialism: Militarism, Humanism, and Occupation.* Montreal: Alert Press, 2010), 12.

141. Hall in Samuel, 1981, 227–40.

142. "Living Conditions." Native American Aid. Accessed May 18, 2011. http://www.nrcprograms.org/site/PageServer?pagename=naa_livingconditions.

Chapter One

Overview

From Pocahontas to Superfink

> *Let's face it: Tonto's a fink. He is like those faithful family retainers who remained loyal to their masters in those after-the-Civil-War pictures. . . . Tonto was an Indian Stepin Fetchit who went around saying "Me white man friend," instead of "Yowzah, boss."*—Ralph Friar and Natasha Friar[1]

This chapter will demonstrate that the few times American Indian characters have been assigned central or key supporting roles in network television programs (fictional dramas), they acted as enforcers of the dominant group's norms. It will also provide an overview of American Indian characters on U.S. network television for the purpose of augmenting the historical record. This overview also demonstrates that there are notable exceptions to Clark's Regulator stage.[2] I use the term "enforcer" to denote a member of a subaltern group who actively works to promote the interests of the dominant group; Clark used the term "Regulator." In other words, he (rarely a she) acts against the interests of his own group's autonomy, sometimes going as far as collaborating with an invading or colonizing force in the manner of Vidkun Quisling of Denmark or Marshall Petain of Vichy France (both are considered notorious traitors who enabled and assisted Nazi invaders). Ralph and Natasha Friar employ the term "fink" (see chapter-opening epigraph) to denote a person who "sells out" his or her group in return for personal advantage.

The Friars seem to define a "fink" as a traitor to his race, that is, an Indian who works for the white man to help whites conquer the territory. Indians themselves sometimes use such a term: "Uncle Tom-Tom," a pun on Harriet Beecher Stowe's slave character Uncle Tom. The Lone Ranger's sidekick,

Tonto, certainly fits this simplistic description—thus the similarly derogatory term "Uncle Tonto." However, in reality, the question of whether a character is or is not a traitor does not lend itself to such a dichotomy. Indians in fact were *never* a homogeneous culture nor have they ever had monolithic interests or traditions; in fact, Native North America contains approximately 250 languages, perhaps as many as 1,000 when the Europeans arrived. Dividing and conquering, for the invaders, was a matter of exploiting existing differences and animosities; many of these peoples were *already* divided, and some welcomed alliances with more technologically advanced outsiders in order to prevail over their traditional enemies. A common tactic among invaders is to look for sympathetic locals to recruit. This was done on several continents. This technique enabled Hernando Cortez with a small regiment of approximately 500 Spanish soldiers—plus the help of thousands of the Aztecs' enemies, the Tlaxcala—to conquer a nation of 250,000. In India and Burma the British recruited sepoys to enforce their laws and exert what they called "indirect control," which was far cheaper—in both money and British lives—than bringing in regulars from the home country. In Rwanda, German colonists recruited the Mwami people to exert indirect rule. And one of the first things the Texas Rangers did in 1835 was to recruit "friendlies" to help fight "hostiles."[3] The point here is that the friendlies were not "selling out" their brethren; they were using the Rangers to help fight their own battles with their own enemies, battles that may have been going on for decades before Anglo-Americans even arrived, so to call these people "finks" is to retrospectively force present-day perspectives onto people in the past. Clark's term "Regulator" is somewhat less loaded. This does not necessarily impose judgment upon the figure, although Clark does say he finds it is ironic that the person enforcing the dominant group's hegemony is from a group that is least likely to benefit from this arrangement.[4]

As many scholars have noted, and as confirmed by this study's surveys, most television representations of American Indians are set in the late 1800s, during the last stages of the conquest of the continent. Perhaps this period is considered the most crucial or the most dramatic (in that it was the most trying or difficult). In the relatively rare instances that American Indians are depicted in the modern era, 58.3 percent of them are portrayed as enforcers of the dominant group's norms—that is, in Clark's term, Regulators (see table 1.1 at the end of this chapter). The survey indicates that the more modern the setting *and* the more prominent the character, the more likely he is to be a Regulator if not an outright collaborator.

Identity, however, whether that of American Indians or Europeans—or any other group, for that matter—is a complex, slippery issue quantitative studies cannot address. This is why interpretive methodologies must also be employed. It is not, as the Friars suggest, as simple a matter as this or that character's being a contemptible collaborator. This Regulator character, although a common stereotype, is nonetheless complex in that he is presented as being like Us and yet somehow *not* like Us—a hybrid of the familiar and the strange. He seems safe for the most part, unthreatening on the surface, yet at the same time he carries an element of danger, insurrection, and this could even conceivably be true of Uncle Tom, the loyal house Negro. There is an underlying tension in the questions the Regulator raises: How loyal is he? Will he remain loyal, or will he turn on us? This tension makes him intrinsically dramatic and entertaining: in order to delight viewers, he dances on the edge of this abyss.[5]

There are numerous other considerations that need to be explored as well. Later chapters analyze these characters in depth, examining such factors as their historical contexts, their political and cultural uses and implications, the constraints and conventions of the television industry itself, genre conventions, gender issues, and cultural meanings carried over from other media such as literature and film (intertextuality).

CLARK'S MODEL APPLIED TO AMERICAN INDIANS

According to Clark, ethnic minorities in the United States generally have had to pass through four stages of literary and visual depictions:

1. *Non-recognition*: A particular minority group is not acknowledged by the dominant group's media even to exist.
2. *Ridicule*: Minority characters are portrayed as stupid, silly, lazy, irrational, or laughable.
3. *Regulation*: Minority characters are presented as enforcers or administrators of the dominant group's norms.
4. *Respect*: This stage is reached when the minority group in question is portrayed no differently than the dominant group and intergroup and interracial relationships are neither notable nor extraordinary.[6]

Clark (now Syed M. Khatib), an African American, was mostly interested in representations of blacks, but he felt his framework might apply to any minority and cited earlier stereotypes of Irish immigrants and Irish Americans as supporting examples.

Clark's framework focuses largely on a specific stereotype: the minority or subaltern enforcer of the dominant group's norms, which he labels Regulator (the *Paris Match* photo analyzed by Barthes in 1957 perfectly illustrates this figure; see the introduction). I tested Clark's framework using American Indian characters to see if it would apply to other minorities and found it had great predictive power, especially in regard to his Regulator stage. However, the term "stages" proved to be problematic in that implies a linear progression; these categories do not progress from one to another but can and do occur randomly.[7] Clark's framework suggests a linear progression, and this may or may not have been true with respect to African American stereotypes; however, this study demonstrates clearly that it is not true in the case of American Indian depictions: there seem to be recurring periods of "backsliding" in which Indians revert from Regulation to Non-recognition.

I compiled a listing of 268 recurring (i.e., appearing in three or more episodes) American Indian characters appearing in regular network-television series from 1949 to 2009. These entries were coded for characteristics such as actors' and characters' names, characters' occupations, and historical settings as well as year of release and length of run (the length of run was an indicator of how successful the program was since this suggests it garnered high enough ratings to be renewed). Several themes or motifs emerged. One is that since the post–World War II era in which U.S. network television emerged, the Good Indian (a variation of the Noble Savage) has been the most prominent representation. There are historical explanations for this phenomenon, mostly having to do with postwar international politics, as discussed in the introduction.[8] The Good Indian usually materializes as an enforcer of the dominant group's norms (Regulator) or sometimes collaborates with Euro-American military forces (as in *Brave Eagle*, 1955, and *Broken Arrow*, 1956). Clark defined "Regulator" as a minority character who works for the dominant power hierarchy, usually for the government. These include military and police officers, judges, schoolteachers, spies, and the like. Of 270 American Indian characters, 31 (11.4 percent) qualified as Regulators. This in itself does not seem significant. However, when American Indians in *contemporary* settings are examined, 58.7 percent were found to be employed in law enforcement, the military, or state-run schools. Still, this is not statistically significant. However, the more *prominent* the character was in the program, the more likely he (rarely a she) was to be a Regulator: when characters (including "half-breed" characters) appeared in starring, costarring, or key supporting (i.e., ensemble) roles, the proportion of Regulators increases to 92.9 percent (i.e., all but one; see table 1.2 at the end of this chapter). Finally, American Indian characters (including half-breeds) in *central or starring* roles number only eight over the course of the six decades; five of these are law-enforcement officers (Cordell Walker, Bravestarr, Nakia Parker, John Hawk, and Sam Buckhart) and the other three (Hondo Lane, Cochise, and Brave Eagle) work for or closely with the U.S. government. *All* of these Indian or half-Indian characters are enforcers of the dominant group's norms—Regulators. Thus the more modern the setting *and* the more prominent the character, the more likely he is to be a Regulator. Only 10 percent of the 270 recurring American Indian characters were women, and among central characters, none are women, so there are gender issues to be addressed as well.

Why so many are enforcers can only be a matter of conjecture at this point. Clark suggested—as indicated in his title "Television and *Social Control*"—that a prime function of minority stereotypes on television was to exert control over minorities themselves.[9] Although this may be a possibility, during the period he addresses, Euro-Americans comprised more than 80 percent of the viewing public, and the target audience for network-television advertising was—and still is—middle-income whites.[10] It is therefore unlikely that network or production executives gave much thought to minorities at all and were more likely interested in appealing to the largest possible audience—and to its prejudices. The 1968 Kerner Commission report, which recommended that the networks include more minority characters in television programs as a potential method of ameliorating racial tension, suggests that network executives indeed had previously given little if any consideration to how minorities were portrayed and perhaps even less to how minorities themselves felt about such portrayals.[11]

While African Americans comprise from 12 to 13 percent of the U.S. population, American Indians, on the other hand, comprise a miniscule market segment.[12] According to a 2000 U.S. census report, American Indians and Alaska natives comprise from 1.5 to 2.5 percent of the population; this figure includes people claiming dual or multiethnic backgrounds. Even in this age of niche marketing, advertisers do not bother appealing to viewerships comprising less than 4 percent of the general population.[13] Thus American Indians are generally overlooked as a viable market for consumer goods. What is more, their viewing habits are not even deemed worthy of documenting by ratings services such as Nielsen and Arbitron.[14]

VIEWER GRATIFICATION AND COMFORTING IMAGES

It is reasonable to assume that the primary function of ethnic stereotypes was to attract or to psychologically benefit—or flatter—members of the dominant group. There is a psychological appeal to this tactic: advertisers hope viewers will, once attracted, associate these benefits with the products being sold. Relatively recent scholarship suggests that one of the benefits (or "gratifications," in modern communications terminology) is to provide *comforting images* for dominant-group viewers.[15] This phenomenon is most likely institutional in nature, intrinsic in the needs of advertisers, most of whom would like their products to be associated with warm feelings—"positive affect"—stimulated by the program's content.[16] This also makes sense on a logical level: Why would viewers regularly tune in to a certain program if it did not provide them some sort of satisfaction? Most mainstream viewers probably prefer programs that reinforce their cultural assumption as well as their prejudices. They probably do not enjoy having their assumptions challenged and therefore would probably not return to a program that does so; it would be a simple matter for them to turn elsewhere, even during the era of the Big Three oligopoly (NBC, CBS, and ABC). The reverse might also be true: if a program fails (i.e., garners low ratings and is therefore canceled), it must be because it did not offer mainstream viewers something they were looking for—presumably a warm, fuzzy feeling. In a sense, this demand for comforting images can be seen as a result of U.S. television's dependence on advertising. Programming—especially during an era when the networks went for the largest audiences rather than niche audiences, as many do today—had to appeal to the prejudices of the larger public or at the very least neither offend nor challenge a large segment of the mass audience's beliefs (this was termed "least objectionable programming"). Programs do not always need to be entirely comforting, however; they may pose challenges to the status quo *so long as the story has a comforting resolution* and returns the situation to a state of normalcy or equilibrium. Network television in the postwar era, then, was inherently conservative in that it tended to support—or at least went out of its way not to challenge—the status quo, because to do so was to alienate viewers and lose advertising contracts.[17]

Another payoff for dominant-group viewers may have been ideological—that of seeing minorities brought under control, thereby reassuring viewers (such as Richard Nixon's "silent majority," discussed in chapter 6) that their own positions in the class and racial hierarchy are secure (see Lippmann epigraph in the introduction).

COLD WAR CONTEXT

It would be impossible to comprehend American Indian representations on U.S. television—or U.S. television in general—without considering the Cold War, for the following reasons:

- The involvement of defense contractors such as General Electric, Westinghouse, and RCA in the development of broadcasting.
- Network television *emerged* during the Cold War—the transcontinental coaxial cable was not completed until 1951.
- The complex postwar international and diplomatic relations related to emerging postcolonial nations, their status within the United Nations, and the United States' image abroad.
- The emergence of the civil rights movement, which greatly affected representations of minorities—or lack thereof—in both film and television, the products of which were closely tied to the U.S.'s international image.
- The explosive proliferation of television westerns during the Eisenhower administration (see figure 0.4 in the introduction).

As noted in the introduction, racial depictions in postwar television were mediated according to international political necessities; these followed a far more benign pattern than had emerged in literature and film.[18] As Aleiss demonstrates, since the dawn of the U.S. film industry, pro-Indian films had been as common as anti-Indian films, if not more so.[19] However, during the interwar years, many anti-Indian films materialized; this approach became considered the standard, what many scholars refer to as the "canonical" western. Examples include the following:

- *The Covered Wagon* (James Cruze, 1923)
- *The Iron Horse* (John Ford, 1924)
- *General Custer at Little Big Horn* (Harry L. Fraser, 1926)
- *Texas Rangers* (King Vidor, 1936)
- Cooper's *Last of the Mohicans* (George B. Seitz, 1936)
- *The Plainsman* (Cecil B. DeMille, 1937)
- *Geronimo* (Paul Sloane, 1939)
- *Stagecoach* (John Ford, 1939)
- *Drums along the Mohawk* (Ford, 1939)
- *Northwest Passage* (Vidor, 1940).

During and after World War II, such films smacked of genocidal proclivities and were either eliminated or severely moderated.[20] There were exceptions, however, such as Ford's *Wagon Master* (1950) and *Arrowhead* (Charles Marquis Warren, 1953).[21]

Television portrayals were an altogether different matter. Televised portrayals of American Indians, for the most part, avoided film's demonic stereotypes.[22] Film's insanely aggressive Indians were largely replaced by Indians as helpless victims who required rescuing by a heroic white man, exemplified in *The Lone Ranger* (which I discuss in detail in chapter 2). The *style* of racism had shifted from outright hatred (and violent repression) to paternalism.

The televisual prototype of the Indian Regulator seems to have emerged during this period in the form of the Lone Ranger's ethnic sidekick, Tonto. His name probably derives from a band of "friendly" Apaches called the Tonto by the Mexicans, which in Spanish means "fool." Tonto, who first appeared on radio in 1933, is an amalgam of several Indian stereotypes, most of which are derived from Defoe, Cooper, and Stowe.[23] *The Lone Ranger*'s stories were simplistic morality plays ostensibly targeted at children, although the program had political overtones that could only have been fathomed by adults.[24] If read in certain ways, the program might have a lot to tell about the postwar international political situation as well as the United States' domestic civil rights struggles: If one were to view the Ranger as the embodiment of Anglo-American law and order and the Indian tribes as metaphors for emerging postcolonial nations, what then was the Ranger's role and what was Tonto's? Tonto becomes a junior partner in the Anglo-American project of expansion, just as *Paris Match*'s "zealous negro" in France (see figure 0.1 in the introduction). In addition, Tonto is very much like a wife to the Ranger, so gender is at issue here as well; as I will show in later chapters, particularly the one on *Broken Arrow* (chapter 3), American Indians are often feminized, depicted in ways white men would rarely if ever be portrayed. This is common in colonialist rhetoric, especially that of the Orientalist type.[25]

The Lone Ranger may have been perceived in its day as having contained an antiracist, pro-Indian theme. However, its intrinsic racism had merely shifted from a blatant to a more subtle form: these less fortunate creatures are treated as friends as long as they participate and assist in the process of Euro-American conquest. This approach, an extension of the paternal metaphor concocted by the Jackson administration, allowed Euro-Americans to have it both ways: they could occupy the land—which, according to the doctrine of Manifest Destiny, was rightfully theirs—and at the same time display a modicum of respect for the peoples being displaced. Thus, as Barthes explains, myth finds a "third way out" of a moral dilemma.[26] Cold War westerns also denied the United States' oppressive past: westerns like *The Lone Ranger* and *Brave Eagle*, treated as if they were quasi-historical documents, "showed" the world that Indians were treated with the utmost respect.

There is another history intertwined with the Cold War: that of the civil rights movement. African Americans had suffered centuries of derogatory stereotypes, from minstrelsy (white comedians performing in "blackface") to an eight-year "blackout" on network television.[27] What implications, if any, might Cold War westerns have articulated with regard to the civil rights movement? Perhaps none, yet this is a question that must be considered. American Indians for the most part did not participate in the civil rights movement. This is because for the most part they did not want to be assimilated or "integrated" into the dominant society. Many if not most prized their separate-but-equal status, a status many African Americans did not. Indians were in fact in danger of losing this distinct status by having their treaties—along with federally funded benefits—terminated.

SURVEY OF AMERICAN INDIAN CHARACTERS: MOSTLY REGULATORS

The Lone Ranger–and–Tonto pairing was the prototype for many more similar figures to come. This section will offer, for the historical record, a compendium of American Indian characters on U.S. television. Literature in the United States has long tried to reconcile or synthesize the two characters, white and Indian, into a viable hybrid that in some ways was intended to represent American culture itself. There had been attempts at merging Indian and white characters in literature as early as 1883 with Francis Bret Harte's story "In the Carquinez Woods" (adapted in film as *The Half Breed* starring Douglas Fairbanks in 1916). However, half-breed characters were usually portrayed either as tragic or as untrustworthy.

In 1953, the native and white characters were successfully merged in the film *Hondo*, produced by and starring John Wayne as half-breed scout Hondo Lane. Lane, however, was identified primarily as white—his "Indian skills" were a bonus. Lane would serve as a model for many superheroic part-Indian characters, the most exaggerated of which would be Cordell Walker of *Walker, Texas Ranger* (CBS, 1993–2001). In 1955, Warner Bros. presented the TV series *Cheyenne* (ABC, 1955–1963), starring Clint Walker as "Cheyenne" Bodie, a white boy raised by the Cheyenne after his parents were killed by Indians (it is not clear whether they were killed by the Cheyenne themselves or by another group; most likely the latter). Much like Daniel Boone, who had been captured by the Shawnee and accepted as one of their own, Bodie is an "Indianized" white man with a vast appreciation for native ways. *Cheyenne* was clearly a pro-Indian text (episode 1, "Mountain Fortress"), presenting an enlightened view—for its time—obviously influenced by Delmer Daves's 1950 film version of *Broken Arrow* (the television version of which debuted in 1956 and will be discussed in chapter 3).

Nearly all American Indian stories are told from the viewpoint of a white protagonist. However, *Brave Eagle* (CBS), making its debut the same year as *Cheyenne* (1955), represented a landmark of sorts: it was "the first TV western told from the Indians' point of view."[28] *Brave Eagle*'s protagonist, Cheyenne chief Brave Eagle (Greg Larsen), is a peaceful man, a friendly who eagerly collaborates with the U.S. Army to defeat "renegade" Indians who oppose Euro-American occupation. Brave Eagle, however, has no idea (in historical terms) what his Euro-American allies have in store for him and his people: for this see Ford's *Cheyenne Autumn* (1964). *Brave Eagle*'s cast of characters includes Brave Eagle's love interest, Morning Star (Kim Winona, a Sioux actor), Brave Eagle's foster son Keena (Keena Nomkeena, a Hopi), and Smokey Joe (Bert Wheeler), a half-breed comic foil. Euro-American characters, mostly army officers, appear sporadically. Most of the program's episodes depict clashes with competing tribes and encroachment of white settlers, from which the army protects these Cheyenne, bringing Pax Americana, the peace-enforcing benefits of imperium (more on this later). Some episodes deal with racial prejudice, with Indians representing minorities in general. *Brave Eagle*'s executive producer was Roy Rogers, one of Hollywood's most famous "singing cowboys," who claimed to be part Indian: Rogers himself was a hybrid, both cowboy *and* Indian.[29]

If Brave Eagle is a collaborator in the project of Anglo-American conquest of the territory, Chiracauha Apache leader Cochise goes a step further: he is an enforcer. A year after *Brave Eagle* made its debut, Cochise (loosely based on a real Apache leader) appeared in a television adaptation of *Broken Arrow* (see details in chapter 3). Both the 1950 film and the ABC-Television series were based on Elliott Arnold's "historical novel" *Blood Brother.*[30] Continuing the *paternal metaphor*, that is, the U.S. government's "protection" of American Indians, federal agent Thomas J. Jeffords serves and protects Apaches in much the same manner Lieutenant John Dunbar would do in *Dances with Wolves* (1990). Michael Hilger calls Cochise "a peace-loving friend to the white man . . . a new version of the Noble Red Man."[31] The issue of Cochise's being a traitor to his own people (or what the Friars call a fink) is openly questioned by other Apache characters in at least two episodes: in "Shadow of Cochise" (season 1, episode 19) the leader of another band, the Coyotera, explicitly calls Cochise "a fool as well as a traitor," and in "Massacre" (season 2, episode 18) an indignant Apache warrior accuses Cochise of "help[ing] the white man in his treachery." However, Cochise is not fazed by these accusations. He, like Brave Eagle, is an intelligent and above all *pragmatic* leader who allies with the side that seems bound to win, just as Quisling or Petain would during World War II. What, then, are this program's political overtones, and what do these say about the historical era—specifically the Cold War—in which they appear? What is their relationship to the civil rights movement?

In 1958, CBS aired *Yancy Derringer* (based on a 1938 pulp novel by Richard Sale). Derringer was a "gentleman adventurer" who was a secretly sanctioned vigilante—a prototypical secret agent—for the city administrator of New Orleans. His assistant was Pahoo-Ka-Ta-Wa (supposedly Pawnee for "wolf who stands in water"), a shotgun-toting, knife-wielding Pawnee who serves as Derringer's bodyguard and strong-arm enforcer. As stereotypes go, Pahoo was even more degrading and atavistic than Tonto, especially since he never said a word. This program lasted one season.[32]

In 1959, NBC brought out *Law of the Plainsman* (also seen in reruns on ABC in 1962), which I analyze in chapter 4. This program represents a landmark in that it offered a clear role reversal from *The Lone Ranger*, which left the air in 1957. The protagonist, an Apache named Sam Buckhart (played by Michael Ansara, who had played Cochise in *Broken Arrow*), is elevated to the position of deputized lawman. It is as if Tonto had been given a promotion: in *Plainsman*, it is the Indian who rides a white horse and wears a badge. Deputy Sheriff Buckhart is a Europeanized Indian, a hybrid combining the best Indian instincts and physicality with Euro-American intellect and respect for the law, an exceptional being who can do almost anything he sets his mind to. Buckhart is not just an ordinary enforcer. Like Superman, he has a sort of secret identity: just as Clark Kent removes his shirt to become

Superman, Buckhart removes his shirt to transform himself, revealing his muscular physique, when he goes after criminals. Buckhart becomes Superfink: stronger than any white man, smarter than any Indian. This program expands on themes initiated in *The Lone Ranger* and *Broken Arrow*, but it also raises issues of masculinity—which had always been a large component of the western—and how the male body is displayed. Certainly the visual treatment of the native male body is much different from the white man's. Another question raised in the program is how closely it follows the government line. Indeed, Twentieth Century-Fox executive Darryl F. Zanuck had met with government officials to offer public-relations assistance.[33] This event must be evaluated in its Cold War context, a time when dealing with foreign (and domestic) threats mattered more than First Amendment principles. *Plainsman* lasted only one season, but it served as a template of sorts for Superfinks such as *Hawk* (ABC, 1966), *Nakia* (ABC, 1974), and, finally, *Walker, Texas Ranger* (CBS, 1993–2001), all of which I explore in later chapters.

Another notable hybrid is half-breed Chief Mingo in the Twentieth Century-Fox coproduction *Daniel Boone*, which I discuss in depth in chapter 5 (NBC, 1964–1970; Mingo was written out in 1967). Mingo (Ed Ames) has a Cherokee mother and an English father (Lord Dunmore, governor of the Virginia colony) as well as an Oxford education. Like Sam Buckhart, he possesses both extraordinary physical and intellectual capabilities. This program reverts to the Cooperesque white hero-Indian sidekick pairing as seen in *The Lone Ranger*. The program's protagonist, Daniel Boone (Fess Parker), is an Indianized white man much like Cooper's Natty Bumppo or *Broken Arrow*'s Tom Jeffords.[34] Mingo, who sides with the Anglo-Americans against the dastardly British, could be construed as yet another collaborator in the Anglo-American conquest; indeed one of the characters in the series, an "Indian princess," calls Mingo a fink. However, this is a complex proposition: as the program shows somewhat accurately, there were Indians on both sides of the conflict. The settlers behave as if they have every right to this territory: this article of faith is simply not questioned by anyone, not even the natives. The series purports to be historical but is a Disney-style sanitized history. This Daniel Boone is presented as a hunter and trapper; the historical Daniel Boone, however, earned a living as a surveyor for land speculators—and did a bit of speculating himself.[35] Thus, as in *The Lone Ranger*, the crucial issues, which are entirely elided, are the acquisition, control, and eventual sale of and profit from the land.

In 1966, ABC-TV presented *Hawk* (which I examine in chapter 6), featuring New York police detective John Hawk, a "full-blooded" Iroquois played by Burt Reynolds, Significantly, he is a one of the few American Indian protagonists whose setting is the contemporary era. Reynolds, who claims to be one quarter Cherokee, had already played several American Indian roles, including half-breed blacksmith Quint Asper on *Gunsmoke* for five years.[36] Detective Hawk is another in a long line of Indian Regulators; he recognizes this and jokingly refers to himself as an "Indian scout." Again, issues of masculinity are raised, with Hawk being a combination police detective, secret agent, and playboy all at once—all were popular representations of masculinity during the postwar/Cold War era. After the runaway success of Britain's James Bond films, which began in 1962 and were unofficially endorsed by President John Kennedy,[37] spies and covert operatives became ubiquitous. It is obvious that even *Hawk* was influenced by the Bond films. As an Indian, however, Hawk is a perennial outsider; this status helps him solve crimes, because he can get information no one else can. This is also true to some extent in the 1974 program *Nakia* (which I analyze in chapter 7).

The Cold War amplified Hollywood's longstanding tradition of glamorizing law enforcement and military officers.[38] In 1967, ABC presented *Garrison's Gorillas* (set during World War II), a sort of precursor to *The A-Team*, featuring four military commandos released from prison on special assignments for the U.S. government. One of these was a stealthy, knife-wielding Apache nicknamed "Chief" (Brendan Boone)—no given name is mentioned—who, like Yancy Derringer's silent sidekick Pahoo-Ka-Ta-Wah, rarely smiles or laughs and is always ready with a knife or gun.[39] According to a novelization of the program that appeared in paperback, "The Chief was of Apache stock. . . . His hands were always busy caressing the handle of the knife sheathed at his waist."[40] Chief is a suave, good-looking youngster (very much like "Kookie" of *77 Sunset Strip*) with an expensive haircut and fashionable clothes. Thus, as in *Hawk*, there is little if any evidence of the character's Indianness—except perhaps in his tracking abilities, physical stamina, and stealth.

In 1967, Warner Bros. and John Wayne (as producer) brought *Hondo* to television (ABC). Several key aspects of *Hondo*'s story line are remarkably similar to *Broken Arrow*'s: Hondo Lane, like Tom Jeffords, respects Apache culture, has lived among Apaches, and was married to an Apache woman who was killed in a U.S. Army massacre (unlike Jeffords, Lane himself is a half-breed). After leaving prison, Lane, like Jeffords, becomes an army scout and an Indian-relations agent, "trying to avoid further bloodshed."[41] Lane becomes close friends with Apache chief Vittorio, an active collaborator who, like *Broken Arrow*'s Cochise, helps enforce the peace treaty by keeping renegade (i.e., noncooperative) Indians in line. This was a short-lived series, lasting only seventeen episodes.

In 1971, Joanie Little Bird (Sandra Ego), a token-minority character, appeared in the rural-Southwest police drama *Cade's County* (CBS) in a minor role as a police dispatcher.[42] The important aspects here are that the Navajo character appears in the contemporary era, and this one is female. She also qualifies as a Regulator. In 1974, ABC again tried the Indian-enforcer format, this time with Navajo police officer Nakia Parker (Robert Forster) in the series *Nakia* (analyzed along with its pilot, a made-for-television movie of the same title, in chapter 7). Parker's issues of identity are what give his story its tension, at least at first: Parker finds his Navajo identity coming into conflict with his job as a Regulator. Parker, a hybrid, can be either Navajo or Anglo as he pleases, but it seems if he wants to continue to be seen and treated as a "regular" person—that is, as white—*he must continually prove his loyalty*, a process that emblematizes the point of this study. In the series pilot, Parker seems to recognize he is a traitor to his people and is troubled by this realization (*Nakia*, 1974). *Nakia* and *Hawk* had much in common: both were produced by Columbia Pictures and appeared on ABC; both characters are Native Americans in their own law-and-order melodramas rather than ethnic sidekicks to white men; both are hybrids combining Indian and Anglo qualities, working for the Euro-American power structure while making efforts to maintain their outsider sensibility, although more so for the rural Parker than for the urbanized Hawk. This "double consciousness" (DuBois's term) apparently makes these officers more sensitive and compassionate, especially toward other outsiders, far more so than the typical, all-business, white cops exemplified by *Dragnet*'s Lieutenant Joe Friday. It also seems to make Indian cops more effective, able to solve crimes ordinary police officers using orthodox methods could not. As David Buxton explains, this was a new breed of police officer, "fighting injustice from above as well as below."[43] They are portrayed not as enforcers but as mediators between the law and the public: they bridge the divide between the state and the public, between young and old. Many of this "new breed" of cops were white, but minority-cop characters fit extremely well into this project.[44]

Also in 1974, Inuit character Abraham Lincoln Imhook served as an ethnic sidekick to a white Alaska state patrolman in the short-lived ABC police drama *Kodiak*. In 1978, Will Sampson (a Creek actor who had played Chief Bromden in *One Flew over the Cuckoo's Nest*) appeared in *Vega$* (ABC, 1978–1981) as Chief Harlon Two-Leaf, "a pony-tailed Native American who helped private investigator Dan Tanna track down criminals amidst the glitter of Las Vegas."[45] In 1985, newsstand operator Billy Night Eyes served as an informant for the police—a fink in the strictest sense—in ABC's *Hollywood Beat*. Again and again we see the Indian character as a Regulator, usually in a subsidiary role to a white savior.

Fueled in part by the film *Dances with Wolves* (1990), the 1990s saw a renewed infatuation with Native Americana and New Age spirituality. David Lynch's iconoclastic series *Twin Peaks* (ABC, 1990–1991) featured police officer Tommy "Hawk" Hill (Michael Horse), mostly notable for his tracking skills. *Renegade* (syndicated, 1992; USA 1996–1997), an action-adventure program, featured bounty hunter Bobby Six Killer (Branscombe Richmond).

In 1993, CBS introduced *Walker, Texas Ranger* (which I analyze in chapter 8). Cordell Walker is the culmination of nearly all televisual Regulators who have gone before—a half-Cherokee hybrid with near-superhuman abilities. He resembles Hondo Lane in that he is a half-breed. Like the Lone Ranger, Walker is very much a superhero; his similarities to Superman, Batman, and the Hulk are remarkable. *Walker* even makes ironic intertextual references to *The Lone Ranger*: both are set in Texas; both feature Texas Rangers; both include a subservient Indian character (in Walker's case, his Uncle Ray); both are squeaky-clean, saintlike figures who (almost always) refuse to kill their adversaries.[46] As in *The Lone Ranger*, *Walker*'s villains are usually Anglo-American, sometimes Mexican. *Walker* has been the only successful (i.e., long-running) program with an American Indian (actually he is only half Indian) in the role of official lawman: *Law of the Plainsman*, *Hawk*, and *Nakia* were all notoriously short-lived. *Walker* did well in the ratings for eight years (1993–2001). *Walker*'s success might have been based on the fact that the character could identify as *either* white or Indian, either establishment figure or outsider, as the case demanded: this ambiguity allowed audiences to read him whichever way they preferred. As Richard Drinnon notes, the white and Indian characters in U.S. literature have been in the process of converging since Natty Bumppo and his real-life progenitor, Daniel Boone.[47] Both characters were Indianized and thereby Americanized. Another issue raised in *Walker* is that of cultural respect and accuracy: one might reasonably expect Chuck Norris, the program's star and executive producer, who claims to be half Cherokee,[48] to care about representing Cherokee culture accurately, yet the program misrepresents Cherokee culture to such an extent that it angers Cherokee scholars.

Star Trek: Voyager (UPN, 1995–2001), set in the future, included a Native American military officer, Commander Chakotay (Robert Beltran). Chakotay's story is similar to Sam Buckhart's (of *Law of the Plainsman*, discussed in chapter 4) in that he rebels against his native upbringing, deciding to become an enforcer of the dominant group's norms; he even acquires an elite education (at the Starfleet Academy). In 2005, *Wolf Lake*, a short-lived horror drama, appeared on CBS, featuring Detective John Kanin (Lou Diamond Phillips) and schoolteacher Sherman Blackstone (Graham Greene). The most recent recurring American Indian characters have been, predictably, police detectives: Chester Lake (Adam Beach), added to NBC's *Law and Order: Special Victims Unit* from 2007 to 2009, Bobby Stillwater (Gregory Cruz) on *Saving Grace*.[49] According to scholars Diana George and Susan Sanders, these programs and others like them tend to place contemporary American Indian characters "back to the level of Tonto, only this Tonto dresses like us and even talks like us."[50] This was clearly true of *Hawk*, *Nakia*, and *Walker*. Yet it cannot be this simple or there would be little dramatic tension: viewers somehow sense that these characters harbor some element of danger—they *appear* to be like us and yet *somehow* different, perhaps having other allegiances.[51] This is what makes them compelling: they appear safe yet at the same time, under the surface, slightly threatening.

TELEVISION PORTRAYALS OTHER THAN AS REGULATORS

Despite having become an anachronism, the pre–World War II Bloodthirsty Savage (i.e., Bad Indian) stereotype occasionally reappeared on TV: for example, the premise of *Wagon Train* (NBC, 1957–1962; ABC 1962–1965) practically demanded the appearance of "hostile Injuns." There are even westerns in which Indians rarely appear: *Gunsmoke*; *Rawhide*; *Have Gun, Will Travel*; and *Bonanza* are largely de-Indianized. American Indians, sometimes hostile, sometimes friendly, occasionally passed through and are minimally acknowledged in these series. In the 1960s, the western genre entered a parodic stage with situation-comedy spoofs such as *F-Troop* (ABC, 1965–1966) and *Rango* (ABC, 1967), in which Indians were cast as comedic figures alongside even more bumbling whites; Milton Berle appeared as an Indian detective, Wise Owl (*F-Troop*, October 6, 1966). Rango, played by Tim Conway, was a Texas Ranger; his sidekick was an Indian scout named Pink Cloud.

In a more serious vein, Cree singer and activist Buffy Sainte-Marie began making semiregular appearances on PBS's *Sesame Street* in 1976 through 1979 (sometimes accompanied by her husband, Sheldon Wolf Child Peters and their son, Dakota Starblanket Wolfchild), with the expressed intention of teaching children that "Indians still exist."[52] American Indian women are a rarity on television. In 1977, Shirley Kirkes appeared as that rarest of rarities, an Indian woman in a contemporary fictional setting, dancer Gladys Little Feather on *Blansky's Beauties* (ABC, 1977), a spin-off of *Happy Days*. In 1983, *The Yellow Rose* (NBC, 1983), a serial set in Texas (reminiscent of *Dallas*), featured a contemporary Indian, John Strongheart, played by the venerable Will Sampson. Both *Blansky's Beauties* and *Yellow Rose* lasted one season.

In the late 1980s and early 1990s, a spate of "revivalist" westerns appeared, such as *Paradise* (later retitled *Guns of Paradise*; CBS, 1988–1991) and *The Young Riders* (ABC, 1989–1992). *Paradise* featured a Paiute character, John Taylor (Dehl Berti), who has been befriended by the white family around whom the story is centered. Like Uncas in Cooper's *Last of the Mohicans* (1985/1826), Taylor is the only surviving member of his nation. Waxing philosophical, Taylor has this discussion with the white children:

> **Taylor:** Indian people left their homes because the white man wanted their land and made them move.

First child: That's not nice!

Taylor: Indians didn't think so, either, so they decided to fight.

Second child: But they lost.

Taylor: You know, the white man is a lot like the giant in your Jack and the Beanstalk. He just takes whatever he wants.

First child: But Jack showed the giant and won!

Taylor: Well, that story took place in England.[53]

This passage begins by admitting that the Euro-American conquest was unjust but ends by dismissing the whole affair as inevitable, minimizing its importance and conveying the idea that it was a foregone conclusion and that there is no point in agonizing over it—it is too far in the past for modern audiences to worry about. George and Sanders suggest that this discursive strategy supplies exactly the sort of "exonerating message" many members of the dominant group need to hear: that the "extinction" of American Indians was not caused by genocidal practices but was part of some indistinct natural process.[54] This is an example of how television comforts and soothes (liberal) audiences who experience occasional pangs of guilt over their forebears' treatment of Indians. Cultural critic Renato Rosaldo labels this "imperialist nostalgia" or "mourning for what one has destroyed."[55] This is worth noting because it tends to recur. It even shows up several years later in *Northern Exposure* ("Things Become Extinct," 1992), as well as in the 1992 film version of *Last of the Mohicans*, in which Chingachgook (played by Russell Means) intones, "The frontier moves with the sun and pushes the Red Man of these wilderness forests in front of it until one day there will be nowhere left. Then our race will be no more. . . . The frontier place is for people like my white son and his woman and their children."[56]

The fairly successful *Dr. Quinn, Medicine Woman* (CBS, 1993–1998) was another revivalist western but with a twist: it featured a white, female protagonist, physician Michaela Quinn (Jane Seymour), who befriends a Cheyenne medicine man, Cloud Dancing (Larry Sellers). This program included several minor Cheyenne characters: some were individualized and had a certain depth of character, yet according to anthropologist S. Elizabeth Bird, these characters have little agency of their own; rather they "function as plot devices to allow Michaela [Quinn] and [her friend] Sully to make a point."[57] Very much like the Cheyenne characters in John Ford's *Cheyenne Autumn* (1964), *Dr. Quinn*'s rather pathetic natives need to be rescued from bad whites by good whites like Quinn and her Indianized boyfriend, Sully, who wears buckskin and speaks fluent Cheyenne.

Northern Exposure (CBS, 1990–1995) is a glaring exception to Clark's Regulator rule: not one of its native characters—Marilyn Whirlwind (Elaine Miles), Ed Chigliak (Darren E. Burrows), Dave (William J. White), Eugene (Earl Quewezance), Uncle Anku (Frank Salsedo), Leonard Quinhagak (Graham Greene), or Lester Haines (Apesanahkwat)—is a Regulator. What is more, these characters are humanized, individualized, and have a sense of humor, which is unusual for Indian representations. *Northern Exposure* distinguished itself by treating Tlingits respectfully—even celebrating them. Still, for the most part they are relegated to the sidelines or depicted as amusing oddities. With the exceptions of Marilyn Whirlwind and Ed Chigliak, Indian characters make only occasional appearances. Whirlwind, a receptionist, and Chigliak, a store clerk, began as less important (in terms of screen time) characters but gained more prominence in later seasons. Whirlwind, a foil for the self-important, New York–bred physician Joel Fleischman, is not given a significant amount of screen time until the third season (1992), when the program suddenly increased its depictions of Tlingit characters—almost as if the producers suddenly realized that these were especially valuable. Half-Tlingit Ed Chigliak is a literal hybrid who struggles with questions of identity, seeking out his biological father. The wealthy Lester Haines, who appears in five episodes (of a total of 110), is an anomaly among Indian characters: he is a crafty, utterly pragmatic and unemotional businessman; the idea of kinship among him and his people is meaningless where there is money to be made. Why *Northern Exposure* takes pains to develop its Indian characters instead of falling back on shopworn stereotypes is an open question. One answer might be that the program is set far from the contiguous forty-eight states, in Alaska, a wild region where normal rules do not apply. Most of the characters, both white and Indian, are eccentric and whimsical, living in a land where logic appears to have little currency or meaning. A nonchalant silliness, which gives the program its distinctive humor, seems to pervade everyday life. Everything is peculiar: dreams and nightmares come to life; as in David Lynch's surreal *Twin Peaks*, pretty much anything can happen. One episode, "Things Become Extinct," does fall back on the Cooperian last-of-his-tribe stereotype; however, Ira Wingfeather (Bryson Liberty) tells Chigliak, who is studying flute making, that the disappearance of Tlingit traditions is "not so tragic. The world used to be full of things that are no longer: mastodons, sabre-toothed tigers, *Indian tribes*, herds of buffalo. Everything gets gone sooner or later. . . . Things become extinct."[58] Wingfeather's vanishing race soliloquy is remarkably similar to John Taylor's in *Paradise* (1990; cited earlier), which in turn draws from *Last of the Mohicans*. What is more, the Tlingit character here notably refers to his people as "things."

There have been few other exceptions to Clark's Regulator rule:

- Zac, a male nurse on *Rachel Gunn, R.N.* (Fox, 1992).
- George Little Fox (Floyd Westerman) on *Dharma and Greg* (ABC, 1997–2002; Westerman left the program in 2001).
- Uncle Ray Walker (Floyd Westerman) on *Walker, Texas Ranger* (CBS, 1993–2001).
- John Redcorn (Jonathan Joss), a massage therapist/New Age healer on animated comedy *King of the Hill* (Fox, 1997–2008).
- Jacob Nighthorse (A. Martinez), a casino operator in *Longmire* (A&E, 2012).
- Henry Standing Bear (Lou Diamond Phillips), a tavern owner, also in *Longmire*.

MINISERIES

As explained earlier, miniseries have been purposely excluded from the data, based on the assumption that, due to these programs' brief life spans, viewers do not become strongly attached to these characters as they might in one that recurs weekly over the period of an entire season (or several, in the case of successful series). Moreover, since these are not intended to become regular programs but are treated as special events, these are allowed to take more risks and thus are exempted from the norms to which regular programs must adhere.[59]

There have been some notable miniseries featuring well-developed American Indian characters. One was *Born to the Wind*, a four-part "summer-replacement" series that aired on NBC in August 1982 (first aired in 1978 as *Indians* on the British Broadcasting Co.). Like 1955's *Brave Eagle*, this was one of those rarest of stories, told entirely from the point of view of American Indians themselves—no white protagonist through whom to relate the tale. It starred Will Sampson as Painted Bear, the leader of an unspecified group of Plains Indians. In 1989, CBS aired a "revivalist" western, *Lonesome Dove*, which began as a Pulitzer-winning novel by Larry McMurtry about two former Texas Rangers. This miniseries spawned three sequels: *Return to Lonesome Dove* (CBS 1993), *Streets of Laredo* (CBS 1995), and *Comanche Moon* (CBS, 2008). It also spawned two syndicated (nonnetwork) series, *Lonesome Dove: The Series* (1994) and *Lonesome Dove: The Outlaw Years* (1995). In June 2005, Turner Network Television presented the six-part epic *Into the West*. Produced by Steven Spielberg, *Into the West* tells the story of the arrival of Euro-Americans by contrasting the viewpoint of a family of white settlers against that of a Lakota family and how these intertwine. The story is centered on a white protagonist who marries a Lakota woman. This is a notable development, sympathetic to Lakota views; still, the program leans heavily on the ostensible inevitability of Euro-American conquest. *Into the West* is told from both Euro-American and American Indian viewpoints.

WHY SO MANY REGULATORS?

The question quantitative research cannot address is, why? What—and whose—purposes do these representations serve? How do they work?

Chiung Hwang Chen observes, "Minorities can only be accepted by the dominant group [and its media] if they fully embrace the values of Euro-American society or become 'virtual whites.'"[60] Robin Means-Coleman and Emily Yochim write that "racial groups only escape condemnation [in the mainstream media] when aligned with the value system of 'Americanized whiteness.'"[61] We have only to look at similar discourses of other colonizing powers, such as imperial Britain or France, to ascertain this pattern. Albert Memmi, a French-speaking Jew who participated in the French colonization of Tunisia (Tunisian Jews tended to align with European invaders against the local Muslims) and later rebelled against it, observed in 1957 that it is not enough for a subaltern people to meekly accept the dominant order:

> In order for the colonizer to be a complete master, it is not enough for him to be so in actual fact, he must also believe in [the system's] legitimacy. In order for that legitimacy to be complete, it is not enough for the colonized to be a slave, *he must accept his role*.[62]

Embracing the norms of Euro-American society is not enough: according to Clark, minority members must go a step further by enforcing them. Just as the slave must constantly demonstrate his gratitude to his master, the minority member of society must risk his life to *prove* his loyalty to the dominant culture; he must "join the team," so to speak. His subjection must appear voluntary. Jean-Paul Sartre wrote in 1943 that the person who dominates another expects his subject to "will its own captivity."[63] The subaltern, then, like the black soldier on the cover of *Paris Match*, is to demonstrate his enthusiasm by taking up the sword in defense of empire. This act endorses the colonial system's legitimacy: it demonstrates to the colonizer that colonial power must be justified because *the subjugated peoples themselves endorse and welcome it*. In similar fashion, the Anglo-American viewing public appears to require simulated endorsements from American Indians:

> These stories, at a mythic level, explain to whites their right to be here and help [them] deal with lingering guilt about the displacement of the native inhabitants—after all, the "good" Indians helped us out and recognized the inevitability of white conquest. . . . *These are people who sided with the invaders*.[64]

In other words, such stories help to alleviate what literary analyst Peter Hulme has called "colonial anxiety."[65]

Table 1.1. Recurring American Indian characters (three or more episodes) set in modern era (58.7 percent are Regulators).

Title (Network, Year)	Character (Actor)	Role	Occupation
Hawk (ABC, 1966)	John Hawk (Burt Reynolds)	Star	Police detective
Garrison's Gorillas (ABC, 1967–1968)	Chief (Brendan Boone)	Costar	Covert operations
Nakia (ABC, 1974)	Nakia Parker (Robert Forster)	Star	Police officer
Kodiak (ABC, 1974)	Abraham L. Imhook (A. Biberman)	Support	Police officer
Vega$ (ABC, 1978–1981)	Harlon Two-Leaf	Support	Informant
Hollywood Beat (ABC, 1985)	Billy Night Eyes	Support	Informant
Twin Peaks (ABC, 1990–1991)	Tommy "Hawk" Hill (Michael Horse)	Minor	Police officer
Bravestarr (syndicated, 1987–1989)	Bravestarr (Pat Fraley)	Star	Federal marshal
Northern Exposure (CBS, 1990–1995)	Marilyn Whirlwind (Elaine Miles)	Support	Receptionist
Northern Exposure (CBS, 1990–1995)	Ed Chigliak (Darren E. Burrows)*	Support	Grocery clerk
Northern Exposure (CBS, 1992–1993)	Leonard Quinhagak (G. Greene)	Minor	Healer
Northern Exposure (CBS, 1994–1995)	Eugene (Earl Quewezance)	Minor	Bartender
Northern Exposure (CBS, 1990–1994)	Dave (William J. White)	Minor	Cook
Northern Exposure (CBS, 1990–1994)	Lester Haines (Apesanahkwat)	Minor	Investor
Rachel Gunn, R.N. (Fox, 1992)	Zac (Bryan Brightcloud)	Support	Nurse
Renegade (USA, 1992–1997)	Bobby Six Killer (B. Richmond)	Support	Bounty hunter
Walker, Texas Ranger (CBS, 1993–2001)	Cordell Walker (Chuck Norris)*	Star	Police detective

Walker, Texas Ranger (CBS, 1993–1994)	Ray Fire Walker (Floyd Westerman)	Support	Ranch hand
Walker, Texas Ranger (CBS, 1995–2000)	White Eagle (Frank Salsedo)	Support	Unknown
Star Trek: Voyager (UPN, 1995–2001)	Commander Chakotay (Robert Beltran)	Support	Military officer
Dharma and Greg (ABC, 1997–2001)	George Little Fox	Minor	Retiree, ghost
King of the Hill (Fox, 1997–2008)	John Redcorn (Jonathan Joss)	Minor	Healer
Wolf Lake (CBS, 2005)	John Kanin	Costar	Police detective
Wolf Lake (CBS, 2001–2002)	Sherman Blackstone (G. Greene)	Support	Schoolteacher
Law and Order: SVU (NBC, 2007–2009)	Chester Lake (Adam Beach)	Support	Police detective
Saving Grace (TNT, 2007–2010)	Bobby Stillwater (Gregory Cruz)	Support	Police detective
Longmire (A&E, 2012)	Officer Mathias (Zahn McClarnon)	Support	Police chief
Longmire (A&E, 2012)	Henry Standing Bear (Lou D. Phillips)	Support	Tavern owner
Longmire (A&E, 2012)	Jacob Nighthorse (A. Martinez)	Support	Casino owner

* Indicates "half-breed" character.

Table 1.2. American Indian characters *in major roles* in any era (82.2 percent are Regulators).

Title (Network, Year)	Character (Actor)	Role	Occupation
Adventures of Pow Wow (synd., 1949)	Unknown	Star	None
The Lone Ranger (ABC, 1949–1957)	Tonto (Jay Silverheels)	Costar	Vigilante
Brave Eagle (CBS, 1955)	Brave Eagle (Keith Larsen)	Star	Indian chief**
Brave Eagle (CBS, 1955)	Morning Star (Kim Winona)	Costar	Indian maiden
Broken Arrow (ABC, 1956–1958)	Cochise (Michael Ansara)	Costar	Indian chief**
Yancy Derringer (CBS, 1958)	Pahoo-Ka-Ta-Wa (X. Brands)	Costar	Bodyguard
Law of the Plainsman (ABC, 1959)	Sam Buckhart (Michael Ansara)	Star	U.S. marshal**
Daniel Boone (NBC, 1964–1970)	Mingo (Ed Ames)	Support	Indian chief**
Hawk (ABC, 1966)	John Hawk (Burt Reynolds)	Star	Police detective
Hondo (1967)	Hondo Lane (Ralph Taeger)*	Star	Army scout**
Hondo (1967)	Vittorio	Support	Indian chief**
Garrison's Gorillas (ABC, 1967–1968)	Chief (Brendan Boone)	Costar	Military officer
Nakia (ABC, 1974)	Nakia Parker (Robert Forster)	Star	Police officer
Bravestarr (syndicated, 1987–1989)	Bravestarr (Pat Fraley)	Star	Federal marshal
Renegade (USA, 1992–1997)	Bobby Six Killer (B. Richmond)	Support	Bounty hunter
Walker, Texas Ranger (CBS, 1993–2001)	Cordell Walker (Chuck Norris)*	Star	Police detective
Longmire (A&E, 2012)	Jacob Nighthorse (A. Martinez)	Support	Casino owner

* Indicates "half-breed" character.

** Indicates active collaborators with U.S. government forces acting against other Natives (35.3 percent).

Table 1.3. Characters in central (starring or costarring) roles *also* set in modern era (100 percent are Regulators).

Title (Network, Year)	**Character (Actor)**	**Role**	**Occupation**
Hawk (ABC, 1966)	John Hawk (Burt Reynolds)	Star	Police detective
Garrison's Gorillas (ABC, 1967–1968)	Chief (Brendan Boone)	Costar	Covert operative
Nakia (ABC, 1974)	Nakia Parker (Robert Forster)	Star	Police officer
Bravestarr (syndicated, 1987–1989)	Bravestarr (Pat Fraley)	Star	Federal marshal
Walker, Texas Ranger (CBS, 1993–2001)	Cordell Walker (Chuck Norris)	Star	Police detective

NOTES

1. Ralph Friar and Natasha Friar, *The Only Good Indian: The Hollywood Gospel* (New York: Drama Book Specialists, 1972), 189.

2. Cedric C. Clark (Syed M. Khatib), "Television and Social Control: Some Observations on the Portrayals of Ethnic Minorities." *Television Quarterly*, Spring 1969, 18–22.

3. Mike Cox, "A Brief History of the Texas Rangers." Texas Ranger Hall of Fame and Museum. http://www.texasranger.org/history/BriefHistory1.htm.

4. Clark, 1969, 18.

5. Indeed, the nagging fear of the native defector's turning on his new masters is based on real experiences. For example, around 1560, several decades before John Smith's arrival in Virginia, a Spanish ship landed in Virginia and picked up a local Pamunkey man, Mangopeesomon, who happened to be Pocahontas's uncle. Mangopeesomon converted to Christianity and traveled to Spain, where he was educated by Jesuits, after which he returned to Virginia to help establish a Jesuit mission at York River. At this point his people nicknamed him "Opechancanough," meaning "one whose soul is white." However, in 1572, he turned on his mentors, leading a massacre against them and destroying the mission. J. Frederick Fausz, "Opechancanough: Indian Resistance Leader." In Gary Nash and David Sweet, eds. *Struggle and Survival in Colonial America* (Berkeley: University of California Press, 1981), 21–37.

6. Clark, 1969, 18.

7. Mary Alice Money identified seven common American Indian stereotypes that she too referred to as "stages." Mary Alice Money, "Broken Arrows: Images of Native Americans in the Popular Western." In Dane Morrison, ed. *American Indian Studies: An Interdisciplinary Approach to Contemporary Issues* (New York: Peter Lang, 1997), 369–88.

8. Mary L. Dudziak, *Cold War Civil Rights: Race and the Image of American Democracy* (Princeton, NJ: Princeton University Press, 2000); Thomas Borstelmann, *The Cold War and the Color Line: American Race Relations in the Global Arena* (Cambridge, MA: Harvard University Press, 2001).

9. Clark, 1969, 22; emphasis added.

10. Richard M. Levine, "Counting Heads: TV Demographics." *APF Reporter* 1(3). Accessed November 26, 2009. http://www.aliciapatterson.org/APF0103/Levine/Levine.html.

11. Kerner Commission, *Report of the National Advisory Commission on Civil Disorders* (Washington: U.S. Government Printing Office, 1968), 389. The commission, on the other hand, could be seen as having proposed the use of television as a means of social control when it made a recommendation that network executives include more minority (especially black) characters, presumably in positive roles. As Clark demonstrated, these inevitably appeared as Regulators, many if not most of which began appearing soon after the report was issued. In other words, the Kerner Commission may have been recommending offering televisual representations as a way of mollifying or co-opting demands for equality—rather than proportional representation in the police force, the black community was offered TV cops instead. Also see Phyl Garland, "Blacks Challenge the Airwaves." *Ebony* 21(1), November 1970, 35–44. Clark/Khatib and I have discussed this issue on several occasions and still do not agree. In any case, his position and mine are not mutually exclusive: social control and positive affect could operate simultaneously.

12. Selene G. Phillips, "'Indians on Our Warpath': World War II Images of American Indians in *Life* Magazine." In Meta G. Carstarphen and John P. Sanchez, eds. *American Indians in the Mass Media* (Norman: University of Oklahoma Press, 2012), 38.

13. Joe Mandese, "In Search of Affluent Viewers." *Television Week* 23(2), January 12, 2004, 47.

14. Hamid Naficy and Teshome H. Gabriel, *Otherness and the Media: The Ethnography of the Imagined and the Imaged* (Langhorne, PA: Harwood Academic Publishers, 1993), 90.

15. Melbourne Cummings, "The Changing Image of the Black Family on Television." *Journal of Popular Culture* 22, March 2004; S. Elizabeth Bird, *Dressing in Feathers: The Construction of the American Indian in American Popular Culture* (Boulder, CO: Westview Press, 1996), 258–59. However, this same concept of *reassurance* for viewers was mentioned by James Baldwin as early as 1963 in his book *The Fire Next Time* (New York: Dial Publishing,

1963), 599, and in 1985 by Mark Crispin Miller, "Deride and Conquer." In Todd Gitlin, ed. *Watching Television: A Pantheon Guide to Popular Culture* (New York: Pantheon Books, 1986), 273.

16. Tony Yeshin, *Advertising* (Florence, KY: Cengage Learning Business Press, 2005), 3, 378; also see Roger Barton, *Media in Advertising* (New York: McGraw-Hill, 1964), 250.

17. This is not true as much in the modern era with more than five hundred cable channels appealing to specialized tastes.

18. Borstelmann, 2001; Dudziak, 2000.

19. Angela Aleiss, *Making the White Man's Indian: Native Americans and Hollywood Movies* (New York: Praeger Publishers, 2005), 30–38.

20. Ibid., 62.

21. Ford later apologized for his disparaging depictions of Indians. Peter Bogdanovich, *John Ford* (Berkeley: University of California Press, 1978), 104.

22. Aleiss, 2005, 60–67; Ken Nolley, "The Representation of Conquest: John Ford and the Hollywood Indian (1939–1964)" in Peter C. Rollins and John E. O'Connor, *Hollywood's Indian: The Portrayal of the Native American in Film* (Lexington, KY: University Press of Kentucky, 1998): 73–90.

23. Most likely all of these began with Defoe's Friday, who appeared in 1719. Cooper, probably borrowing from Defoe, paired a white hero and an Indian sidekick: Natty Bumppo and Chingachgook. Stowe's Uncle Tom was likely derived, at least partly, from Friday as well; Stowe was a great admirer of Defoe's work. James H. Foster, *The Rungless Ladder: Harriet Beecher Stowe and New England Puritanism* (New York: Cooper Square Publishers, 1970), 30. Fiedler, however, asserts that Uncle Tom was derived from Pocahontas. Leslie Fiedler, *The Return of the Vanishing American* (New York: Stein & Day, 1968), 17. As to Friday, I would speculate it is very likely that Defoe based this character at least partly on the legend of Pocahontas, who was already famous as Rebecca Rolfe, wife of Englishman John Rolfe. She was brought to London in 1614, where she was treated as a celebrity. Kiros Auld, "Pocahontas." Powhatan Museum of Indigenous Arts and Culture. http://www.powhatanmuseum.com/Pocahontas.html; Linwood Custalow and Angela L. Daniel, *The True Story of Pocahontas* (Golden, CO: Fulcrum Publishing, 2007), 58–59, 79–81. However, Hulme suggests that the idea may have been borrowed from Shakespeare's play *The Tempest* (1610) in the relationship between Prospero and Caliban. Peter Hulme, *Colonial Encounters: Europe and the Native Caribbean* (London: Metheun, 1986), 195, 202, 220.

24. Mary E. Bickel, *George W. Trendle: Creator and Producer of the Lone Ranger, the Green Hornet, Sgt. Preston of the Yukon, the American Agent, and Other Successes* (New York: Exposition Press, 1971), 120.

25. Edward Said, *Orientalism* (New York: Vintage Books, 1979), 3.

26. Roland Barthes, *Mythologies* (London: Paladin, 1972), 129.

27. Melvin P. Ely, *The Adventures of Amos 'n' Andy: A Social History of an American Phenomenon* (New York: Free Press, 1991), 228.

28. "Brave Eagle, 1955." Turner Classic Movies. Accessed April 25, 2009. http://turnerclassic.moviesunlimited.com/product.asp?sku=D28770.

29. Richard Severo, "Roy Rogers, Singing Cowboy, Dies at 86." *New York Times*, July 7, 1998, A1.

30. Elliott Arnold, *Blood Brother* (New York: Duell, Sloan & Pearce, 1947).

31. Michael Hilger, *The American Indian in Film* (Metuchen, NJ: Scarecrow Press, 1986), 98.

32. The secret-agent craze had not yet hit the United States in 1958. A similar series, *The Wild Wild West*, described by its creator as "James Bond on horseback," ran on CBS from 1965 to 1969 but without the Indian sidekick. Quoted in Susan King, "In Secret Service of Fun." *Los Angeles Times*, November 4, 2008. http://articles.latimes.com/ 2008/nov/04/entertainment/et-wildwildwest4.

33. Jackson Committee, "March Meeting Notes." April 10, 1953. Washburn Abbott Papers, Box 7. Dwight D. Eisenhower Library, Abilene, Kansas.

34. Bumppo was probably based on the real Daniel Boone. Frederick Zackel, "Robinson Crusoe and the Ethnic Sidekick." *Bright Lights Film Journal* 58, November 2007. http://

www.brightlightsfilm.com/30/crusoe1.html; also see K. Randel Jones, *In the Footsteps of Daniel Boone* (Winston-Salem: John F. Blair, 2005), 13.

35. Jones, 2005, 32.

36. Burt Reynolds, *My Life* (New York: Hyperion, 1994), 103, 114, 129, 136.

37. David Buxton, *From "The Avengers" to "Miami Vice": Form and Ideology in Television* (Manchester: Manchester University Press, 1990), 108.

38. Tricia Jenkins writes, "The Federal Bureau of Investigation established its entertainment office in the 1930s and has used its influence to bolster its image of the bureau in radio programs, films and television shows such as *G-Men* (1935), *The Untouchable* (1959–1963), *The FBI Story* (1950), and *The FBI* (1979)." Tricia Jenkins, "Get Smart: A Look at the Current Relationship between Hollywood and the CIA." *Historical Journal of Film, Radio and Television* 29(2), June 2009, 229. Philip D. Smith and Kristin Natalier, *Understanding Criminal Justice: A Sociological Perspective* (Thousand Oaks, CA: Sage Publications, 2004), 161; William Marling, "Detective Fiction on Radio, TV, and in Comics." Detnovel.com. http://www.detnovel.com/OtherMedia.html; Lawrence Suid, *Sailing on the Silver Screen: Hollywood and the U.S. Navy* (Annapolis, MD: Naval Institute Press, 1996); Lawrence Suid, *Guts and Glory: The Making of the American Military Image in Film* (Lexington: University Press of Kentucky, 2002).

39. Tim Brooks and Earl Marsh, *The Complete Directory to Prime Time Network and Cable Television Shows* (New York: Random House, 2003), 448; Jack Pearl, *Garrison's Gorillas* (New York: Dell Books, 1967), 15, 80–85.

40. Pearl, 1967, 15.

41. Brooks and Marsh, 2003, 548–49; "Hondo: Episode Guide." TV.com. Accessed April 24, 2009. http://www.tv.com/Hondo/show/5143/episode.html.

42. Brooks and Marsh, 2003, 180–81; "Native Americans." TVAcres.com. Accessed May 1, 2008. http://www.tvacres.com/ethnic_native_american_AK.htm.

43. Buxton, 1990, 122.

44. Perhaps the best example of this "new breed" of cop was ABC's 1968 series *The Mod Squad*, featuring a trio consisting of a young, hip white guy, an idealistic white "hippie" female, and a black-nationalist radical.

45. "Native Americans." TVAcres.com.

46. The Lone Ranger actually does lethally shoot a villain who is taking aim at him in the 1956 film *The Lone Ranger*. Standards for violence were less stringent in film than in television. Directed by Stuart Heisler (Wrather Productions/Warner Bros. Pictures, 1956).

47. Richard Drinnon, *Facing West: The Metaphysics of Indian-Hating and Empire-Building* (Minneapolis: University of Minnesota Press, 1980), 22.

48. Ira Berkow, "At Dinner with Chuck Norris." *New York Times*, May 12, 1993, C1.

49. Since this chapter was written, three more have appeared on the 2012 police series *Longmire* (A&E), set in Wyoming. Only one of these three Indian characters could be considered a Regulator: Officer Mathias (Zahn McClarnon), chief of the Cheyenne reservation's police. The others are Henry Standing Bear (Lou Diamond Phillips), a bar owner, and Jacob Nighthorse, who operates the Cheyenne's casino.

50. Diana George and Susan Sanders, "Reconstructing Tonto: Cultural Formations and American Indians in 1990s Television Fiction." *Cultural Studies* 9(3), 1995, 448.

51. In this sense they are reminiscent of characters from the 1950s who might have been aliens or even communists. On the issue of Indians as quasi-communists, see Philip French, *Westerns* (New York: Viking Press, 1973), 80–82.

52. "Buffy-Sainte Marie" (biography). CreativeNative.com. 2003. http://www.creative-native.com/biograp.htm.

53. From a 1990 episode, quoted in George and Sanders, 1995, 435.

54. George and Sanders, 1995, 436–41.

55. Renato Rosaldo, *Culture and Truth: The Remaking of Social Analysis* (Boston: Beacon Press, 1993), 69.

56. *Last of the Mohicans: Director's Expanded Edition*, directed by Michael Mann (Morgan Creek Productions/ Twentieth Century-Fox Films, 1999). DVD.

57. Bird, 1996, 251.

58. *Northern Exposure.* Season 3, episode 13, "Things Become Extinct." (Cine-Nevada Productions/Brand-Falsey Productions/MCA Television, aired January 20, 1992). This soliloquy bears a remarkable resemblance to one purportedly made in 1855 by Chief Seattle of the Duwamish, which intoned, "tribe follows tribe, and nations follow nations, like the waves of the sea. It is the order of nature." "Speech of Chief Seattle." In John Bryant, ed., *American Literature* (Boston: Pearson Custom Publications, 2003), 304.

59. This tradition was established with the surprise success of the ABC-TV adaptation of Alex Haley's *Roots* in 1977.

60. Chiung Hwang Chen, *Mormon and Asian-American Model-Minority Discourses in News and Popular Magazines* (Lewiston, NY: Edwin Mellen Press, 2004), 209.

61. Robin Means-Coleman and Emily Yochim, "Symbolic Annihilation." In Wolfgang Donsbach, ed. *The International Encyclopedia of Communication* (Malden, MA: Blackwell Publishing, 2008), 4923.

62. Albert Memmi, *The Colonizer and the Colonized* (New York: Orion Press, 1991), 89; emphasis added.

63. Jean-Paul Sartre, *Being and Nothingness* (New York: Washington Square Press, 1966), 367.

64. Bird, 1996, 2–3; emphasis added.

65. Hulme, 1986, 2.

Chapter Two

Cold War Avengers

The Lone Ranger and Tonto in The Lone Ranger *(1949–1957)*

> *God has not been preparing the English-speaking and Teutonic peoples for a thousand years for nothing. . . . He made us master organizers of the world to establish system where chaos reigned. . . . He has made us adept at government that we may administer government among savage and senile peoples. . . . He has marked the American people as His chosen nation to finally lead in the redemption of the world.*—Senator Albert Beveridge[1]

The Lone Ranger was hardly a novel character when he was developed by Detroit radio-station owner George Trendle in 1933. Before purchasing WXYZ-AM, Trendle had been a cinema owner/manager for many years and knew what kind of characters boys liked.[2] The radio program came to television in 1949, during the early stages of the Cold War.[3] This chapter examines two primary issues contained in *The Lone Ranger*: what the Ranger represents as "white savior" and what his ethnic sidekick Tonto represents as his assistant or "apprentice white man."[4] Being white here has far deeper implications than just skin color: the Ranger's Anglo-Saxon-Protestant values say a lot about the culture as well as the program's ideological aims. Tonto's presence raises the question of whether, and if so, how, a minority figure—or "colored man"—can fit into this project.

BACKGROUND OF THE PROGRAM

The Lone Ranger was one of the earliest network-television programs recorded on film.[5] Trendle hired former MGM producer Jack Chertok to produce the TV series, with George B. Seitz Jr. as director. The format of the program, which aired on ABC until 1957, was, in an industrial sense, a continuation of the B-movie western that had been traditional fare in U.S. cinemas.[6]

The Lone Ranger's introduction offers a fairly intricate, fast-paced bit of camera work. The rousing horns of Rossini's "William Tell Overture" blare as a medium tracking shot, taken from a vehicle running alongside the Ranger, pulls back to a full shot of him and his white horse at full gallop. The frame allows a bit of the landscape to be seen in the background: it is pristine, comprising blue sky, rolling hills, and lots of trees. There are no buildings, telephone poles, wires, or modern edifices. In fact, there are no signs of civilization at all.[7] This lack of human habitation is crucial and will be discussed later.

The western hero must, as an Indianized white man, be connected with the wilderness.[8] This is often accomplished by associating him with mountains. The Ranger's head appears to be connected to the mountains (figure 2.1), implying strength and stability. He wears a pistol in a metal-tipped holster on his right hip, which illustrates his connection to technology. The

Figure 2.1. Clayton Moore in opening sequence of *The Lone Ranger*, 1949.

Ranger appears to be chasing someone. This is confirmed as he pulls the gun from its holster and fires at an unseen target ahead of him, presumably escaping villains. The pistol, as well as his white horse, gleams against the dark trees in the background (figure 2.1). Suddenly the rider abandons the chase and takes a diversion up a hill to get the lay of the land.

At the top of the hill, he guides his huge white horse into a standing maneuver, rearing back on its hind legs in a show of strength (figure 2.2). Series star Clayton Moore was an accomplished horseman, and this was one of the reasons the producers selected him for the role.[9] Issues related to archetypal images of horse and man are too complex to go into here, but suffice to say that the horse and control of it are symbolic of the rider's mastery over nature.[10] A low-angle shot provides a spectacular view of both horse and rider, creating a heroic image of the character, framed against the sky, with which he has a special relationship (figure 2.2). The rock formation in the upper-left area of the screen, the most powerful region, has symbolic significance—the rock, like the mountains, implies strength, stability, steadfastness; together, the Ranger and the horse fuse with this symbol of nature and eternity. Again, the hero is visually conflated with the wilderness and

Figure 2.2. Moore demonstrates horsemanship skills near Ranger Rock.

mountains (or rather a rock formation).[11] At the same time he is also associated with the sky, from whence he came.

MESSIANIC TROPES

At issue here is the construction of the white savior and how he is depicted as divine. That religious tropes appear throughout the program is not surprising considering that its executive producer, George Washington Trendle, once aspired to become a minister.[12] Reverend Rob Archey describes both the radio and television series as Bible lessons sans Scripture.[13] Trendle himself announced in a promotional flier that the Ranger was "the embodiment of a granted prayer."[14] Besides that, in 1949 the United States was fighting an ongoing ideological battle with the "godless communists" of the Soviet Union: U.S. leaders sought to differentiate the nation from the USSR by celebrating the United States' deep religiosity, a point later underscored by the 1954 insertion of the phrase "under God" into the Pledge of Allegiance.

This program contains several specific tropes that identify the Ranger as godlike. For one, he experiences a symbolic death and rebirth, a common theme among dying-and-regenerating gods. For another, he is always inexplicably clean: there is never a speck of dirt on him, and he never needs a shave, even after he has been unconscious for days. His custom trapeze-artist outfit is always immaculate, even after he has dragged himself through dirt and rubble.[15] The Ranger's hat, too, is immaculately white. The western hero is always inexplicably clean.[16] This obsession with cleanliness and grooming has a Puritan resonance: "dirt is Other."[17] Moreover, cleanliness is associated with respectability.[18] The white horse is a well-known symbol that lets the viewer know the rider is godlike. For example, in the New Testament book of Revelation, Christ is described as descending from the heavens on a white horse.[19] Lancelot of Arthurian legend also rode a white horse. There is a knight on a white horse in Edmund Spenser's sixteenth-century poem "Sonnet XIV," based on the book of Revelation.[20] Allen Kerkeslager writes that the white horse as a symbol of divinity is derived from the myth of the sun god Apollo.[21] "The Man on the White Horse" is also an important trope in the Mormon Church: founder Joseph Smith alluded to a rider who arrives when civil upheaval is at a peak and the "Constitution is hanging by a thread," rescuing the nation at the last minute.[22] Ariel Dorfman notes, "the superhero's arrival is always *providential.*"[23]

As he sits atop his steed, a medium shot shows the Ranger scanning the horizon from his left to his right, head framed against a bright blue sky, with clouds serving as a sort of halo (figure 2.3). The sky, serving as a frame for his portrait, suggests that he came from the heavens. Richard Dyer observes, "Men [framed] against the horizon are a cliché of aspirational propaganda."[24]

Figure 2.3. The Ranger scans the horizon in a watchful, godlike pose. Here he is visually associated with the clouds.

Besides being all-seeing, his gaze seems proprietary, as if he were a king surveying his domain. Martin Pumphrey writes, "Looking is established as a form of [masculine] control."[25] A narrator underscores this suggestion of divinity, intoning:

> This is a story of one of the most mysterious characters to appear in the early days of the West. He was a fabulous individual—a man whose presence brought fear to the lawless and hope to those who wanted to make this frontier land their home.[26]

THE LAND AND ITS VALORIZATION

The land and how it is portrayed is a crucial issue in *The Lone Ranger*, as it is in most westerns. After all, the land is the prize: who controls it and how it is to be used is generally the primary issue. *The Lone Ranger*'s exterior scenes were shot at the Iverson Ranch in Chatsworth, California, near San Fernando. Like any western, this series features lots of panoramic shots that emphasize the landscape. Many if not most television westerns open with long vistas valorizing the land. Since wide vistas are not considered particularly effective on the small screen in creating an impressive spectacle or an awe-inspiring gaze, one might reasonably assume they are included for some other reason. These genre conventions come from film and are used to quickly identify these programs as westerns: man on horse plus long shot of land equals western. This trope was probably popularized by seminal director John Ford: Ford always made sure to accentuate the grandeur of the landscape in his westerns, often shooting in picturesque Monument Valley.[27] The beauty of the landscape is part of the myth of the winning of the West: in the canonical western, control of the land is what whites and Indians fight over. The land here is like a beautiful woman, inviting the male gaze. Each side wants to possess it: "a 'virgin' land is implicitly available for defloration and fecundation."[28]

How *The Lone Ranger* deals visually (and narratively) with issues related to the land is telling. In episode 1 ("Enter the Ranger"), a typical long shot shows pioneers in covered wagons chopping wood and struggling to improvise shelters. These scenes are viewed through a parchment map labeled "Southwest Territory of the US" with a hole burned in the center, through which the camera records the action. This could be intended to provide a sort of evidentiary veracity. Significantly, there are no Indians, only Conestoga wagons cascading down hillsides from the east. This gives a historically inaccurate impression, to say the least: before the arrival of Euro-Americans, Texas had been one of the most densely populated American Indian regions on the continent.[29] Dorfman remarks, "In this version of history the Indians are not the former possessors of the soil. . . . The rights to that land are established first by historically destroying its initial possessors and later suppressing that act of destruction."[30] The land is simply shown as vacant and inviting, and this view, juxtaposed with the announcer's declamation, serves to foster a crucial element of justificational ideology: the land was uninhabited and therefore available for the taking.[31] This (mis)portrayal of the land as vacant is a variation of the sixteenth- and seventeenth-century concept of *terra nullius* ("empty land"). This bit of legal sophistry had often justified the taking of the land by Euro-Americans because, ostensibly, no one was making good use of it.[32] However, the commonly held assumption that American Indians were generally nomadic hunter-gatherers is erroneous.[33]

At first glance, the western genre's "predilection for grandiose views of scenic panoramas and expansive vistas" appears to suggest a sort of rustic egalitarianism, equated in the popular mind with the yeoman farmer.[34] The "valorization" of the land, however, implies much more than a natural adoration found in landscapes in European art à la Brueghel. What we are seeing in westerns is a specific type of *reification* in which the land takes on a mystical, godlike character. This tradition had been established in the United

States by painters such as Thomas Cole in the 1830s. Landscape paintings took on distinct ideological functions. "Artists began to equate the country's unspoiled wilderness with the new nation's seemingly limitless potential."[35] These landscape artists mystified—and myth-ified—the land, equating it with the Promised Land of the Old Testament.[36] Virginia Wright Wexman writes, "The implication of such visions of vast territories embraced in a single possessive gaze is far from egalitarian; they rather suggest the more hierarchical, dominating dimension of American imperialist aspirations."[37] The land, in Wexman's view, is personified as female, which coincidentally or not correlates to the traditional view of the Earth derived from paleolithic fertility religions.[38] Wexman suggests that this longing gaze at the land is comparable to the "male gaze" Laura Mulvey described in 1975.[39] It looks upon the land in a proprietary manner as a successful, bourgeois male might gaze upon a female nude: the viewer *fantasizes* about owning and controlling the body but settles for a representation.

The struggle for control of the land and how it should be treated have been key issues in the clash between Europeans and Native Americans since the Pilgrims arrived in 1620 to create Plymouth Colony. The winning-of-the-West myth is the story of wresting the land from its inhabitants. Churchill states, "Land is the absolutely essential issue defining viable conceptions of Native America. . . . Contests for control of territory underlie the virtually uninterrupted and ongoing pattern of genocide suffered by American Indians."[40] When Euro-Americans arrived in North America, many if not most native groups had no concept of ownership of land; they thought of the land as a commons available for all to use as necessary.[41] Euro-Americans, on the other hand, not only had an elaborate system of ownership, surveys, and deeds, but the Pilgrims (i.e., Puritan separatists) also had a vision of the land as their own special gift from God in which His chosen people would be granted "dominion" over the earth and all its inhabitants and were encouraged to "subdue it."[42] The Pilgrims preached that they were God's chosen people and that He had given them this land.[43] As far as they were concerned, this was *their* home, *their* promised land, to which they were, in a spiritual sense, returning. Sociologist Ghassan Hage explains, "The national 'home' fantasy is . . . of one of the oldest fantasies of domestication and is primarily an ecological fantasy of domesticated nature: the Garden of Eden."[44] The fundamentalist longing, then, is to re-create and return to Eden.[45]

This clash over the land and how it should be treated reaches into the cosmology of both races. The conflict between adherents of the ancient feminine principle ("mother earth") and the masculine ("sky father") reflects a primal struggle that goes back to ancient fertility religions. In what is ethnocentrically considered the more modern of these two views, the masculine element was awarded dominance by its own—notably male—deity.[46] Anglo-

Americans wanted to bend nature to make it fit them. Historian Christopher Lasch remarks that the dream of subjugating nature has a narcissistic element: "It seeks to restore the primal illusion of omnipotence."[47] It appeared to Anglo-Americans as if the aboriginals were letting the land—and its valuable resources—go to waste. American Indians, however, felt it was their place to fit themselves to nature rather than the other way around.[48] The natives, believing that the land like the air belongs to all, at first saw little problem in sharing it with the settlers—not realizing that they themselves were to be excluded and would have to fight to keep from being forcibly removed.

Yet when *The Lone Ranger*'s narrator states that Anglo settlers "had to fight," the fight is not with Indians *but with evil white men* brandishing six-shooters, terrorizing Euro-American settlers. Marauding Indian villains of the prewar canonical western, such as Ford's *Stagecoach*, were generally replaced by marauding white villains.[49] Due to international political complexities (discussed in chapter 1), images of overt conflict between white settlers and American Indians tended to create diplomatic difficulties between the United States and emerging postcolonial nations. There are few clashes with Indians in this series and these are resolved after a few skirmishes leading to reserves being granted to Indians by the U.S. government and protected by the army. The few episodes that involve conflicts with local Indians depict them as fierce warriors taking reasonable steps to guard their reserved lands.[50]

The villains on *The Lone Ranger* are always white men, even though a Texas Rangers historical site unequivocally states the organization was founded *to fight Indians.*[51] When the narrator says, "an organization was formed to combat the evil forces," we see a gang of marauding white bandits rather than marauding Indians. The Rangers here are in search of the notorious Cavendish gang, led by Butch Cavendish, who had been robbing and terrorizing settlers. Captain Dan Reid's unnamed younger brother (the future Lone Ranger), is among these six men, but he is not singled out; at this point he is no one special as far as the viewer is concerned. The six Rangers are led into an ambush by a "half-breed scout" (in the words of the narrator), a stranger named Collins. The viewer is alerted to the fact that Collins has native ancestry by his fancy buckskin jacket and the large, protruding feather in the band of his black hat. The half-breed, neither fish nor fowl, is singled out for a special brand of ignominy; like the Oriental, he is congenitally treacherous, not to be trusted by anyone, neither the Rangers nor the Cavendish gang, as ringleader Cavendish himself later points out.

A close-up shot reveals Collins's facial hair and unkempt appearance—a scraggly beard growth—indicating he is of questionable character. In contrast to the slovenly Collins, the Texas Rangers are fastidiously clean. True to his nature, Collins has led them into a trap. One Ranger is hit—the sound of a shot is heard as he grabs his shoulder—and he immediately keels over on top of the lifeless body of his fallen comrade. Later, Cavendish shoots Collins, explaining that if Collins would betray the Rangers, he would do the same to the gang members.

In the next sequence the Ranger, left for dead, pulls himself, facedown, through the dirt and rubble with one arm dragging behind him in an attempt to reach water. This should have ripped his outfit to shreds, but like Superman's costume, it remains intact, indestructible. Collins also revives and crawls up to a ledge on a mountain face.

ENTER THE REGULATOR

Next we see a long shot of another lone figure carrying a sidearm, emerging through a pass in the canyon on a small, spotted horse. This is the Ranger's ethnic sidekick, Tonto. In the original radio series, Tonto did not appear in early episodes, until the program's writers decided a sidekick would allow the Ranger someone to explain things to.[52] However, on the television program, Tonto was an integral character from the start. He subsequently became the quintessential stoic Indian, which in some ways was a positive step and in other ways locked Indians into a sort of hapless, slow-witted stereotype. Author Sherman Alexie, a Spokane, said Tonto "was the first really mainstream, pop-culture Indian figure, the monosyllabic stoic-Indian type."[53] In a *Los Angeles Times* op-ed piece, Alexie wrote, "I hated Tonto then, and I hate him now."[54]

Tonto's initial appearance contrasts with the Ranger's in that he is first shown in a long shot, making him appear tiny; moreover, he is hemmed in on all sides by huge cliffs that make him seem even more puny, merely part of the landscape (figure 2.4). The spotted horse surely has some significance: in western movies the "painted pony" (*pinto* in Spanish) is a common trope of Indianness.[55] However, if the Ranger's white horse echoes his whiteness, then Tonto's mixed-colored horse in this instance indicates this rider's hybridity or *partially* white nature. Tonto wears a buckskin jacket like Collins's, also signifying his Indianness. Unlike half-breed Collins, however, this rider is superclean and clean-shaven, so in Puritan terms he is trustworthy. His hair, although long, is carefully parted, slicked back (tucked under in the rear in a sort of bun) and neatly fastened in place with a thin, black headband, another signifier of Indianness. His hair is long, suggesting femininity, yet it is also very tidy, signifying goodness. A soft focus is used in this close-up (figure 2.4), a technique normally reserved for women to give them an ethereal "glow."[56] His clothes, though made of buckskin, are fastidiously neat and tidy. The rider comes across the near-lifeless body of the Ranger by the side of a spring. Looking for some identification, he pulls

Figure 2.4. Close-up of Tonto peering down at the Ranger; focus is "soft."

out a chain from the Ranger's shirt, as if it were an army dog tag. An extreme close-up shows that the chain holds a ring with a figure of a Phoenix-like bird on its face. A close-up of Tonto's startled face shows he recognizes it: "You Kimo Sabe!" Tonto exclaims.[57]

Suddenly there's a flashback, to about twenty years previous: a medium-shot of the young Ranger holding a young Indian's nearly lifeless body. There are burning tipis and smoke in the background. Tonto narrates: Decades ago, the adolescent who would become the Lone Ranger had run across an unconscious Indian boy in the aftermath of his tribe's being massacred by renegade Indians, whom we never see. Tonto appears to be the last of his tribe, a trope emanating from Cooper's *Last of the Mohicans*. The Lone Ranger and Tonto become brothers of a sort, the Ranger being the older, wiser brother. This is emblematic of their continued relationship. In a montage of two-shots, the young Reid, who becomes the Ranger, nurses the Indian boy back to health. Reid gives the Indian boy a horse. Another two-shot shows him handing Tonto the reins, then a close-up shows Tonto slipping a ring on Reid's right hand in a gesture reminiscent of a wedding ceremony.[58] Tonto explains he could not accept the horse as a gift without giving a gift in return. An extreme close-up shows the future Ranger's remarkably clean, freshly manicured hand sporting the ring. At this stage the two characters become like husband and wife; this is reinforced in a later scene in which Tonto washes the Ranger's hat. This scenario probably derives from Defoe's *Robinson Crusoe*, in which Friday functions as a combination manservant and wife to Crusoe.[59]

Flashing forward, back to the canyon where the Ranger was ambushed, the Ranger and Tonto's dialogue is seen in two-shots—but the Ranger's head is turned away from the camera, his face obscured. The Ranger now recalls the Indian's name—Tonto, which, coincidentally, happens to be Spanish for "fool" or "dolt." "Now me take care of you," Tonto replies in his pidgin English. This situation too probably derives from Defoe's tale, in which Crusoe rescues Friday from a rival group of Caribs who intend to eat him; consequently Friday dedicates his life, which would have been forfeited anyway, to serving Crusoe.[60] A montage of two-shots shows Tonto nursing the Ranger back to health, feeding him, and washing him with a damp cloth. Realizing his life is still in danger from the ambushers, the Ranger decides to elude them by faking his own death, but, as Tonto points out, he needs a disguise. "From now on, I'll wear a mask." Using a knife, Tonto cuts a perfect mask out of leather taken from the Ranger's dead brother's vest. At this point, neither the Ranger nor Tonto have any family, friends, or social ties: they are free to fight crime full-time. It is a convention in western stories that the hero be a mysterious stranger with no job, no family, no friends.[61]

Tonto hands the Ranger his hat, now gleaming, explaining, "Me wash in stream; dry in sun—make whiter." This gesture is like a christening. The Ranger ceremoniously dons the hat; he is now ready to fight crime everywhere. Thus, despite his constant valorization of the law, Reid concludes he can better enforce the law by not being constrained by it: he fakes his death in order to become a masked vigilante. That all this is illegal is elided. Nonetheless, the program's law-and-order rhetoric flows continuously. In another dialogic two-shot, the Ranger again cites the sanctity of the law. Tonto responds, "Me want law here too—for all." Such a statement presumes that the only legitimate law is Anglo-American jurisprudence.[62] Only Anglo-American law can bring peace in a Hobbesian world, as the flashback of the destruction of Tonto's band exemplifies. This exchange between rescuer and hapless victim—who eagerly becomes a junior partner in the enterprise—fits squarely into colonial discourse, as Said describes it: "What could be better for [the colonizer's] self-image than native subjects who express assent to the outsider's knowledge and power, implicitly accepting European [in this case Euro-American] judgment on the undeveloped, backward, or degenerate nature of their own society?"[63]

Back to the present: the villainous Collins now awakens. As the Ranger and Tonto leave the canyon, Collins lies in wait with a rifle pointed in their direction. The Ranger decides to capture Collins. A shot from Collins's point of view high up in a mountain crevasse shows the Ranger and Tonto running across the canyon floor toward him. His point of view resembles that of a hawk watching scurrying mice.

Collins wears a rag wrapped around his head, presumably to cover the bullet wound Cavendish gave him. The effect is that of an Arabic headdress, and with his dark hair, beard growth, and flowing buckskins, he looks vaguely eastern, exotic, and threatening. Low-angle shots then follow Tonto and the Ranger climbing the mountain's rocky face to sneak up on Collins. As the Ranger scuffles up the cliff side in his thick riding boots, Tonto nimbly climbs upward in what appear to be orthopedic shoes.

An extreme close-up of the Ranger's knee-high boots, complete with silver spurs, shows him losing his footing. This is not just a close-up used to build suspense; the boot is a *synecdoche*: not only does it stand for the Ranger himself but for justice (Texas Rangers) and modernity. This rugged boot requires not only craftsmanship but technology (for the metal buckles and spurs) to build. It also symbolizes strength: this boot can step on those who defy its owner. The camera slowly tilts upward from the Ranger's tenuous fingerhold on the precipice to Tonto worriedly looking down from the top of the mountain (figure 2.6). Tonto is now framed against the sky in a godlike pose, just as the Ranger was in the introduction. After agreeing to join the Ranger in his divine quest for justice, Tonto takes on some of the trappings of divinity. The camera continues its upward tilt toward the clear blue sky, to remind the viewer where the Lone Ranger came from: the heavens. What is more, the spectator/viewer here is being positioned as God: all-seeing, all-knowing (i.e., omniscient point of view).[64]

MACRO-LEVEL READINGS

It is advantageous to interpret *The Lone Ranger* television program and its cultural significance on multiple levels. To do so requires situating the series in its contemporaneous political, cultural, and historical context. At face value, this extremely nationalistic program celebrates the building of Anglo-American civilization, constructed through acts of law enforcement. Law enforcement is presented as the very foundation of U.S. civilization—despite the fact that the Ranger himself is engaged in numerous illegal activities. The Ranger is a vigilante, motivated by retribution, a popular attraction for U.S. television viewers and moviegoers.[65] However, he is not looking for the murderers of his brother and his compadres but decides to fight evildoers wherever he may find them. Despite constantly citing the sanctity of the law, the Ranger decides he can perform this task better by not being constrained by it; by faking his own death, he becomes a shadow, answerable to no law.

Figure 2.5. Tonto, framed against the sky, peers down at the Ranger.

Since such a reading makes little sense, it is reasonable to assume that the program has some deeper ideological meaning.

A second reading might view the program as a quasi-historical crime series offering a *commentary* on law enforcement as it was conducted in the 1880s compared to how it was conducted in 1949. The implication is that if old-fashioned "western justice" dispensed by self-appointed vigilantes was good enough for our forebears, it ought be good enough for us now. Since the Ranger himself violates the law in order to enforce it, the underlying message might be that the now-ineffectual police forces in the United States should ignore the law whenever there is an extraordinary threat to a peace-loving populace. Indeed, political economist Michael Parenti asserts that this has been the message of most crime shows on U.S. television, adding that films and television programs have historically propagated images and ideologies that are supportive of authoritarianism, violence, and vigilantism:

> [M]edia dramas teach us that . . . the ills of society are caused by individual malefactors and not by the economic system, and that *the police should be given a freer hand* in combating the criminal element, using generous applica-

> tions of violence without [paying] too much attention to [citizens'] constitutional rights.[66]

Crime dramas often suggest that the democratic system simply does not work; therefore, it is up to righteous individuals to step in—even if they have to break the law themselves—to honor it. This has been a durable theme in U.S. literature, film, and television, and it is certainly true of *The Lone Ranger*.[67]

A third reading might present the Ranger as an allegorical stand-in for the president of the United States or perhaps an idealization of what a president should be and how he should act—in other words, a fantasy substitute. This is inherently a political statement and an endorsement of authoritarianism. *The Lone Ranger* as a presidential figure offers a "hardline" alternative to—or example for—incumbent Harry S. Truman. Although Truman played the role of hardheaded zealot—for example, he claimed he had "never lost a minute of sleep" over his decision to drop atomic weapons on civilian targets in Japan[68]—he was not nearly aggressive enough for the hard-line militarists of the Cold War. In 1949, the year *The Lone Ranger* made its television debut, Republicans—and even some Democrats—blamed Truman for "losing" China to a communist revolution. By the 1952 presidential election it appeared Truman might "lose" Korea, too. The president had been constrained by the United Nations charter, which was ratified by the U.S. Senate in 1945, making it the law of the land. Thus the Korean War was, by law, to be fought under the auspices of the UN, not by the United States. The commander of the UN forces was General Douglas MacArthur, whose plan involved the use of dozens of atomic bombs in Korea and China.[69] However, British and French officials strenuously objected to what they saw as MacArthur's reckless ideas. On the other hand, U.S. hard-liners charged that Truman's decision to rule out the use of nuclear weapons was spineless and soft and that he had caved in to the UN. Many felt the United States should act unilaterally, ignoring the objections of our allies.

In this reading, the Ranger symbolizes a leader very much like MacArthur. The program's villains might have represented renegade leaders such as Josef Stalin, Mao Zedong, Marshal Tito, Mohammad Mossadegh, Gamal Nasser, and Ho Chi Minh.[70] It is illuminating to look at the series in light of developments in the Cold War as well as the anticommunist witch hunts, which were just beginning in 1949. By episode 3 ("The Lone Ranger's Triumph"), it becomes clear that the gang members have infiltrated local government in the fictional town of Colby. Cavendish's foot soldiers have installed themselves in the second and third most important positions in local law enforcement, as sheriff's deputies, without the sheriff's knowing their true motives. The sheriff—an allegorical allusion to Truman, perhaps—is well intentioned yet slow and fails to see what is going on right under his nose. Only the Ranger understands the true extent of these threats and how to deal with them. This scenario might be an allusion to communist infiltration of the executive branch of the federal government—possibly a reference to Alger Hiss, a state-department official who had been accused of being a Soviet spy in 1948 and was tried for perjury in 1949, the same year the Ranger made his television debut.

A fourth reading could posit the Ranger as the personification of the United States itself. Trendle's biographer, Mary Bickel, suggested exactly this.[71] In this reading, the villains might symbolize "rogue" nations, and the dead Texas Rangers could be a metaphor for failed international law—the UN perhaps—suggesting that the Lone Ranger, that is, the United States, must act unilaterally, imposing a Pax Americana.

Ultimately, however, the Lone Ranger figure should be examined as a metonym for benevolent white supremacy. This "white man's burden," bringing the Anglo-Saxon "gift for governing" to the barbarians of the world, has been a recurring theme in literature, used to justify conquest.[72] In this fantasy, helpless Indians eagerly welcome the white man's superior law and order.[73] Tonto and his people, decimated by renegade Indians, desperately need to be rescued by a white savior. Tonto enthusiastically welcomes the imposition of Anglo-American hegemony: "Me want law here too—for all." He is eager to fight on the side of the noble white savior because he believes in Anglo-American law and order and is painfully aware of the inherent inefficacy of American Indian oral tradition as opposed to the Anglo-American written law (this same motivation recurs in the 1959–1960 ABC program *Law of the Plainsman*, discussed in detail in chapter 4). Tonto's is a Hobbesian world where Native Americans require protection from each other. Ghassan Hage notes, "Left to themselves, 'ethnic' cultures are imagined as unable to coexist. It is only the white effort to inject 'peaceful coexistence' that allows them to do so."[74] Tonto abandons his own way of life and his own people to join the Ranger in his quest to establish Pax Americana, becoming an apprentice white man, an enforcer of the dominant group's norms, a Regulator. Tonto and the Ranger agree to align in order to fight a mutual enemy, with Tonto as junior partner. This scene asks Cold War viewers to believe that Indians *asked* Euro-Americans—as Tonto does—to come into their territories and impose their laws. Tonto *invites* Anglo-Americans, personified by the Ranger, to intervene.[75]

Westerns often made allegorical references to foreign policy. Film scholar Stanley Corkin writes, "[T]he western was well suited to convey important ideological rationales for postwar US foreign policy, including the inevitability of American expansion and the strategies for hegemony that guided the Truman administration's foreign policy."[76] If the philosophy of *The Lone Ranger* is any indication, it is apparent that the series' creator, George Trendle, though a staunch Republican, agreed to some extent with Truman on the issue of interventionism: the Ranger is clearly an interventionist. The difference is that Trendle felt the United States should act unilaterally, whereas Truman was constrained by the United States' treaty with the UN. Tonto is presented as pragmatic, having seen firsthand the weakness in his people's lack of preparedness and defensive technology. This could have been a comment on the debate over military "preparedness" (or lack thereof) that was used by U.S. leaders to sell the Cold War to Congress and the public. This program also demonstrates that the cooperation of "colored" (i.e., nonwhite) citizens is crucial in the fight against "godless communism," personified by white evildoers.

Issues of masculinity abound. Not only is Tonto feminized, so are his people: decimated by a stronger tribe, they are helpless. Can it be coincidence that a young, virile white man materializes to fulfill the rescue mission? In the United States, being a savior is almost always a masculine proposition: the white knight comes riding to someone's rescue, usually a "damsel in distress." According to Brenda M. Boyle, to be in need of rescuing is to be feminine.[77] Countries too can be seen as either masculine or feminine.[78] Both Korea and South Vietnam were feminized in the popular (i.e., mediated) view, in need of being rescued,[79] much like Tonto.

TRENDLE, FREEMANSONRY, AND WHITE SUPREMACY

One reason Trendle admired the Texas Rangers may have been that the outfit had been founded and staffed by Scottish Rite Freemasons, and Trendle himself was one.[80] During the period *The Lone Ranger* was aired (1949–1957), Scottish Rite promoted the most extreme sort of racist views. The point here is that Trendle, as an active member of Scottish Rite, was steeped in these views. For example, an excerpt from Scottish Rite's official publication, the *New Age Magazine*, published during *The Lone Ranger*'s first season, declared, "The hand of Providence has chosen the Nordic people to bring and unfold the new order of the world. . . . Providence has chosen the Nordic people because they have prepared themselves and have chosen God."[81] Belief in Nordic racial superiority did not originate in Germany: remarkably similar beliefs had been in circulation in England and in the United States (i.e., Anglo-Saxonism) before Germany emerged as a nation.[82] According to Reginald Horsman, Anglo-Saxons have long believed they have a "gift for governing," which they have a duty to bring to the rest of the world, whether or not it is welcome.[83]

Where, then, does the American Indian fit into this worldview? In *The Lone Ranger*, Tonto serves as the Indians' representative; he welcomes the white savior on their behalf. In turn he is accepted into the Anglo-Saxon-Nordic project if—and only if—he is willing to assist in this project of

Anglo-American control of the land. Tonto becomes an apprentice white man, a Regulator, doing the dirty work for the white man.[84] It might also be illuminating to ask, where do African Americans fit into this vision? The simple answer is they do not. Not only are blacks not included in Trendle's vision of the Old West—even as *third*-class citizens—they simply do not exist. They have been, in Clark's term, relegated to "Non-recognition"—or, as Gerbner and Gross would say, "symbolically annihilated."

TONTO TROPES IN POPULAR CULTURE

Tonto thus serves as a symbolic ambassador for American Indians. He becomes an apprentice white man, like Defoe's Friday, assisting in the imposition of Anglo-American rule—in Clark's term, a Regulator. Tonto has defected, gone over to the other side. Thus "the one who [in reality] is plundered becomes [in this fantasy] a guardian of the laws that gave rise to that plunder."[85] The Friars employed a more vehement term: "Tonto's a fink"[86] Dakota scholar Vine Deloria Jr. also uses the term "fink" to describe Tonto and others like him:

> Tonto was everything the white man always wanted the Indian to be. He was a little slower, a little dumber, had much less vocabulary, and rode a darker horse. . . . Like the Negro butler and the Oriental gardener, Tonto represented a silent, subservient subspecies of Anglo-Saxon [*sic*] whose duty was to do the bidding of the all-wise white hero.[87]

Tonto is unable to think for himself, nearly always dependent on instructions from the white savior. Indeed, Tonto was so obsequious—almost a manservant or valet—that this became a source of humor for actors Jay Silverheels (Tonto) and Clayton Moore (the Ranger): Tonto's standard reply to the Ranger's commands, "Me do!" became a private joke between the two actors.[88]

As the conformity and nationalism of the 1950s gave way to the dissent and rebellion of the 1960s, Tonto's relationship with the Ranger would be unmercifully mocked in pop culture. There have been several famous jokes satirizing Tonto's servility, implying that it was unlikely, disingenuous, and ironic, that the joke was really on the paternalistic white man. In 1958 *Mad* magazine presented a two-page panel by cartoonist Joe Orlando:

> **Ranger:** Indians! Indians all around us! Well, Tonto, ol' Kimo Sabe, it looks like we're finished.
>
> **Tonto:** [drily] What you mean, "we"?[89]

Later versions of the joke substituted, "What you mean, 'we,' white man?" This joke was immortalized in song by Oscar Brown Jr.[90] *The Lone Ranger* became a target for comedians, especially Lenny Bruce, who even made jokes about Tonto's being sodomized by the Ranger.[91] Bill Cosby also satirized the absurdity of Tonto's subservience, noting that the last time Tonto went into town, he was brutally attacked by settlers (Tonto's beating and near lynching occurred in the 1956 film version of *The Lone Ranger*, and in *The Lone Ranger and the Lost City of Gold* [1958], Tonto gets into a scuffle with a local sheriff, who shoots him). Cosby's comic routine illustrates what he feels would have been a more rational—and likely—response:

> **Ranger:** Tonto, you go to town.
>
> **Tonto:** You go to hell, Kimo Sabe!
>
> **Ranger:** I want you to get the information.
>
> **Tonto:** Information say Tonto not go to town. That what information say.[92]

In 1969, Silverheels himself satirized his own role, on NBC's *Tonight* with Johnny Carson. Carson plays a career counselor helping Tonto find another job after the cancellation of *The Lone Ranger*. In pidgin English, Tonto confides that he and the Lone Ranger had had a falling out after "thirty lousy years" because "him find out what Kimo Sabe really means."[93] Even decades after *The Lone Ranger* left ABC, entertainers were still satirizing the servile relationship between Tonto and the Ranger. A 1986 song by Lyle Lovett titled "If I Had a Boat" included the following verse:

> The mystery masked man was smart
> He got himself a Tonto
> 'Cause Tonto did the dirty work for free
> But Tonto he was smarter
> And one day said, "Kimo sabe, kiss my ass
> I bought a boat, I'm going out to sea"[94]

These jokes play on the tension created by the suspicion—or outright expectation—that the native enforcer will one day come to his senses, realize he is being exploited, and turn on his alien master, much like the sepoys of India did in 1857. The common expectation in these jokes is that the enforcer finally, inevitably turns on his master. Of course there is every reason for the master to expect his slave to turn on him. This expectation was clearly noted by Robinson Crusoe:

> I [had] no doubt that if Friday could get back to his own nation again, he would not only forget all his religion but all his obligation to me and would be

> forward enough to give his countrymen an account of me and come back with perhaps a hundred or two of them and make a feast upon me.[95]

CONCLUSION

Despite its unexamined and apparently unremarkable racist overtones, *The Lone Ranger* was seen as progressive in its day. It was never a canonical western in which Euro-Americans are seen waging all-out exterminationist warfare against American Indians (and vice versa) as in, for example, Ford's *Stagecoach*. As early as the 1933 radio series, Indians in *The Lone Ranger* were portrayed as friendly and their relations with whites genial.[96] *The Lone Ranger* television version was "politically correct" for its time, and this might account for a measure of its success. Nonetheless—aside from the obvious fact that African Americans were excluded—it is inherently racist in that natives are seen as unable to survive without assistance from Euro-American protectors. This is a recurrent theme that reappeared, for example, in Delmer Daves's 1950 film *Broken Arrow* (as well as in the television version, discussed in chapter 3), in David Lean's *Lawrence of Arabia* (1963), and in Kevin Costner's film *Dances with Wolves* (1990). It also resurfaced in 2009 in James Cameron's *Avatar*. A comment from Vine Deloria Jr., written forty years before *Avatar*, sounds as if it could have been written about *Avatar*'s Na'vi people, who are actually thinly veiled representations of American Indians: "So rigid was the stereotype of the childlike Indian that all efforts by [them] to come together were passed off as the prattling of children who could not possibly do anything without instructions from their white friends."[97] In *Broken Arrow*, the white savior is Tom Jeffords of the Bureau of Indian Affairs, and the American Indian figure is Cochise of the Chokonen Apaches. In *Broken Arrow* the Apache takes on some of the characteristics of the Euro-American savior figure and becomes more hybridized. The white man, too, is a hybrid, becoming notably Indianized.

NOTES

1. Senator Albert J. Beveridge, address to U.S. Senate, January 9, 1900. *Congressional Record*, 56th Congress, 1st session, pp. 704–12. Quoted in David Erwin Harrell, Edwin S. Gaustad, and John B. Boles, *Unto a Good Land: A History of the American People*. Volume 2, *From 1865* (Cambridge: Wm. B. Eerdmans, 2005), 743.

2. Mary E. Bickel, *George W. Trendle: Creator and Producer of the Lone Ranger, the Green Hornet, Sgt. Preston of the Yukon, the American Agent, and Other Successes* (New York: Exposition Press, 1971), 197.

3. The Ranger also appeared in various media, such as a series of eighteen paperback novels from 1936 to 1956 as well as two feature films released through Republic Pictures in 1938 and 1939. He also appeared in a weekly comic strip syndicated through King Features from 1938 to 1971, and in 145 comic books published by Dell Comics from 1948 to 1962.

4. Frederick Zackel, "Robinson Crusoe and the Ethnic Sidekick." *Bright Lights Film Journal* 58, November 2007. http://www.brightlightsfilm.com/30/crusoe1.html.

5. The first program filmed specifically for network TV was *Your Show Time* on NBC, aired January 21, 1949. *The Lone Ranger* debuted on ABC eight months later, September 15, 1950.

6. It was relatively easy to convert movie production to television production, especially with westerns, because Hollywood's B-movie mills were already set up for this sort of product, already had sets and costumes on hand, and had already worked out elaborate and efficient manufacturing procedures.

7. It was shot at Iverson Ranch near Chatsworth, CA.

8. Will Wright, *Six Guns and Society* (Berkeley: University of California Press, 1977), 58.

9. Moore had recently starred in Republic Pictures' feature film *Ghost of Zorro* (1949). The Ranger character was partly modeled after Zorro. Clayton Moore with Frank Thomas, *I Was That Masked Man* (New York: Taylor Trade, 1998), 113.

10. Tyler Parker, "The Horse: Totem Animal of Male Power." In Gerald Mast and Marshall Cohen, eds. *Film Theory and Criticism* (Oxford: Oxford University Press, 1974), 590–99.

11. Wright, 1977, 58, 82.

12. Bickel, 1971, 15.

13. Joe Southern, "The Lone Ranger a True Believer," *The Silver Bullet*, n.d. http://www.series.net/ranger/index/html. Retrieved February 20, 2008.

14. David Wilson Parker, "A Descriptive Analysis of the Lone Ranger as a Form of Popular Art." Unpublished PhD dissertation, Northwestern University, 1956, 212.

15. Moore himself had been a trapeze artist in his youth. Moore, 1998, 31–34. His Ranger costume was designed by Nudka "Nudie" Cohen of Las Vegas, who created western-themed outfits for the likes of Gene Autrey, Roy Rogers, Robert Redford, Elvis Presley, Porter Wagoner, Buck Owens, Gram Parsons, and Elton John. Jamie Lee Nudie and Mary Lynn Cabrall, *Nudie: The Rodeo Tailor* (New York: Gibbs Smith, 2004), 110.

16. Martin Pumphrey, "Why Do Cowboys Wear Hats in the Bath? Style Politics for the Older Man." *Critical Quarterly* 31, October 1989, 78–100. Reprinted in Douglas Pye and Ian Cameron, eds. *The Movie Book of the Western* (London: Cassel, 1996), 51–53.

17. "Puritanism—the Philosophy of the English Revolution." Department of English, Umea University, Umea, Sweden. Accessed May 25, 2011. http://www.eng.umu.se/culturec/PUR.htm.

18. Suellen Hoy, *Chasing Dirt: The American Pursuit of Cleanliness* (Oxford: Oxford University Press, 1995), 4–5.

19. Revelation 6:2, New American Standard Bible (Grand Rapids, MI: Zondervan, 2011).

20. Edmund Spenser, "Sonnet XIV." In Robert Atwan and Laurance Wieder, eds. *Chapters into Verse: A Selection of Poetry in English Inspired by the Bible from Genesis through Revelation* (Oxford: Oxford University Press, 2000), 448 .

21. Allen Kerkeslager, "Apollo, Greco-Roman Prophecy, and the Rider on the White Horse in Revelation 6:2." *Journal of Biblical Literature* 112(1), Spring 1993, 116–21; also see Wayne Jackson, *Select Studies in the Book of Revelation* (Stockton, CA: Courier Publications, 1995), 13–14.

22. Robert W. Smith and Elisabeth A. Smith, eds., *Scriptural and Secular Prophecies Pertaining to the Last Days Including Prophecies by Members of the Church of Jesus Christ of Latter Day Saints* (Whitefish, MT: Kessinger Publishing, 2004).

23. Ariel Dorfman, *The Emperor's Old Clothes: What the Lone Ranger, Babar, and Other Heroes Do to Our Minds* (Durham, NC: Duke University Press, 2010), 85; emphasis added.

24. Richard Dyer, *White* (London: Routledge, 1997), 152.

25. Pumphrey, 1996, 58.

26. *Enter the Ranger: The First Three Episodes*. VHS. (Apex Films/UAV Video, 1997; first aired September 15, 1949).

27. Deborah A. Carmichael, *The Landscape of Hollywood Westerns: Eco-criticism in an American Film Genre* (Salt Lake City: University of Utah Press, 2006), 7, 212; Gerald Peary and Jenny Lefcourt, *John Ford: Interviews* (Jackson: University Press of Mississippi, 2001), 56.

28. Ella Shohat, "Gender and the Culture of Empire." In Hamid Naficy and Teshone Gabriel, eds. *Otherness and the Media: An Ethnography of the Imagined and the Imaged* (Langhorne, PA: Harwood Academic Publishers, 1993), 47.

29. Ward Churchill, *Struggle for the Land: Native North American Resistance to Genocide, Ecocide, and Colonization* (San Francisco: City Lights Books, 2002), 332.

30. Dorfman, 2010, 67.

31. However, there are later episodes in which American Indians and their presence—as well as their right to occupy certain "reserved" lands—are freely acknowledged: episode 6, "War Horse" (air date October 20, 1949); episode 8, "Renegades" (November 3, 1949); episode 34, "Rifles and Renegades" (May 4, 1950); and episode 44, "White Man's Magic" (July 13, 1950). As in Delmer Daves's film *Broken Arrow* (1950), the Ranger is respectful of Indians and their (limited) rights, working diligently to preserve peace between the races.

32. Vine Deloria Jr., *Custer Died for Your Sins: An Indian Manifesto* (Norman: University of Oklahoma Press, 1988), 10. In 1629, Puritan leader John Winthrop extended the concept, calling it *vacuum domicilium*. John Winthrop, "General Considerations for the Plantation in New England," 1629, quoted in Francis Jennings, *The Invasion of America: Indians, Colonialism and the Cant of Conquest* (Chapel Hill: University of North Carolina Press, 1975), 71, 135–38.

33. Lyman Carrier, *The Beginnings of Agriculture in America* (New York: McGraw-Hill, 1923), 41. Gerard Fowke also demonstrated that Indians had long been outstanding farmers and agriculturalists. Gerard Fowke, *Archeological History of Ohio: The Mound Builders and Later Indians* (Columbus: Ohio State Archaeological and Historical Society, 1902), 478, 480; cited in Richard Drinnon, *Keeper of Concentration Camps: Dillon S. Myer and American Racism* (Berkeley: University of California Press, 1987), 22–23.

34. Virginia Wright Wexman, "The Family on the Land: Race and Nationhood in Silent Westerns." In Daniel Bernardi, ed. *The Birth of Whiteness: Race and the Emergence of U.S. Cinema* (New Brunswick, NJ: Rutgers University Press, 1996), 142.

35. "Landscape Painting." National Gallery of Art. 2011. http://www.nga.gov/education/american/landscape.shtm.

36. David Bjeljac, *American Art: A Cultural History* (New York: Harry N. Abrams, 2001), 195.

37. Wexman in Bernardi, 1996, 129.

38. Bettina L. Knapp, *Women, Myth, and the Feminine Principle* (Albany: State University of New York Press, 1997), 12–13.

39. Laura Mulvey, "Visual Pleasure and Narrative Cinema." *Screen* 16(3), Autumn 1975, 6–18.

40. Ward Churchill, *Fantasies of the Master Race: Literature, Cinema, and the Colonization of American Indians* (San Francisco: City Lights Books, 1998), 131.

41. Eli M. Salzberger, "Economic Analysis of the Public Domain." In Lucie Guibault and P. Bernt Hugenholtz, eds. *The Future of Public Domain: Identifying the Commons in Information Law* (Alphen aan den Rijn, Netherlands: Kluwer Law International, 2006), 34. Also see Lewis Hyde, *The Gift: Imagination and the Erotic Life of Property* (New York: Vintage, 2007), xix. Also see Carlos L. Rodriguez, Craig S. Galbraith, and Curt H. Stiles, "American Indian Collectivism: Past Myth, Present Reality." *PERC Reports* 24(2). Property and Environment Research Center. Summer 2006. http://www.perc.org/articles/article802.php.

42. Genesis 1:28, New American Standard Bible (Grand Rapids, MI: Zondervan, 2011). This is often referred to as the "dominionist" view. Also see Jeanne Kay, "Human Dominion over Nature in the Hebrew Bible." *Annals of the Association of American Geographers* 79(2), June 1989, 214–32.

43. Reginald Horsman, *Race and Imperialism* (Cambridge, MA: Harvard University Press, 1981), 3, 82–83; Conrad Cherry, *God's New Israel: Religious Interpretations of American Destiny* (Englewood Cliffs, NJ: Prentice-Hall, 1971), 26.

44. Ghassan Hage, *White Nation: Fantasies of White Supremacy in a Multicultural Society* (Annandale, NSW: Pluto Press, 1998), 167.

45. Kay, 1989, 214.

46. Franceska Perot, "The Re-emergence of the Divine Feminine and Its Significance for Spiritual, Psychological and Evolutionary Growth." Unpublished PhD dissertation, Westbrook University School of Psychology, College of Humanities, 2004, 23.

47. Christopher Lasch, *The Culture of Narcissism: American Life in an Age of Diminishing Expectations* (New York: W. W. Norton, 1979), 244.

48. Vine Deloria, 1988, 180–89.

49. Norman A. Scotch, "The Vanishing Villains of Television." *Phylon* 21(1), Fall–Winter, 1960, 58–62.

50. In the episode titled "War Horse" (air date October 20, 1949), the story line is rather similar to that of the film *Broken Arrow*, in which the Indians are provoked by white criminals, which transfers the blame for conflict between the races to a few deviant individuals, while the U.S. Army is depicted as protectors of the peace. It is entirely possible that this episode was inspired or at least influenced by Elliott Arnold's 1947 novel *Blood Brother*, the source for the movie *Broken Arrow* (1950).

51. Mike Cox, "A Brief History of the Texas Rangers." Texas Ranger Hall of Fame and Museum. http://www.texasranger.org/history/BriefHistory1.htm.

52. Robert Siegel, "The Lone Ranger: Justice from Outside the Law." National Public Radio. January 17, 2008. http://www.npr.org/templates/story/story.php?storyId=18195323.

53. Gretchen Bataille, ed., *Native American Representations: First Encounters, Distorted Images, and Literary Appropriations* (Lincoln: University of Nebraska Press, 2001), 9.

54. Sherman Alexie, "I Hated Tonto (Still Do)." *Los Angeles Times*, June 28, 1998. http://articles.latimes.com/ 1998/jun/28/entertainment/ca-64216.

55. "History of the Breed." American Paint Horse Association. Accessed May 1, 2008. http://www.apha.com/breed/history.html.

56. Jeff Jones, *How to Photograph and Edit Women* (Raleigh, NC: Lulu.com, 2006), 71; Graeme Turner, *Film as Social Practice* (London: Routledge, 2000), 92.

57. "Kimo Sabe" is supposedly Potawatamie for "trusty scout." Moore, 1998, 121. Actually it was the name of a campground owned by the father-in-law of James Jewell, one of the early radio show's writers. David Rothel, *Who Was That Masked Man?* (New York: A. S. Barnes, 1981), 40.

58. This brotherhood ritual could have been influenced by Elliott Arnold's 1947 novel *Blood Brother*, upon which the 1950 film *Broken Arrow* was based.

59. As Peter Hulme notes, Crusoe and Friday live in a state of "domestic bliss," and their relationship is "tinged with erotic delight." Hulme, *Colonial Encounters: Europe and the Native Caribbean* (London: Metheun, 1986), 212. Defoe himself, in the voice of Crusoe, wrote, "[T]he three years in which we lived [on the island] together [were] perfectly and completely happy." Daniel Defoe, *Robinson Crusoe* (London: CRW Publishing, 2007), 267.

60. Defoe, 2007, 245. Given this context, however, *The Lone Ranger*'s backstory oddly declines to address the Ranger's and Tonto's decades-long separation. If Tonto is to dedicate his life to serving the Ranger, how and why did they become separated, and why does Tonto decide decades after the fact to make good on his debt to the Ranger? My purpose here is to illustrate that these are merely clichés strung together from bits and pieces of colonialist discourse, some of which go back to the eighteenth century if not further.

61. Wright, 1977, 129.

62. Most American Indian groups had codes of conduct but not written laws. However, the Cherokees had adopted a written legal code by 1808. The Cherokees also had their own law-enforcement system, including marshals, sheriffs, and constabularies. Bob L. Blackburn, "From Blood Revenge to the Lighthorsemen: Evolution of Law-Enforcement Institutions among the Five Civilized Tribes in 1861." *American Indian Law Review* 8(1), 1980, 53.

63. Edward Said, *Culture and Imperialism* (New York: Vintage Books, 1993), 149.

64. The viewer's identification with the cinematic apparatus is deemed primary. Jacques Aumont et al., 1992, 214.

65. Drucilla Cornell, *Clint Eastwood and Issues of American Masculinity* (New York: Fordham University Press, 2009), 73–75.

66. Michael Parenti, *Make-Believe Media: The Politics of Entertainment* (New York: St. Martin's Press, 1992), 2–3; emphasis added.

67. John Shelton Lawrence and Robert Jewett, "Interplay between Popular Myth and American Foreign Policy." Paper presented at the American Culture Association, New Orleans, LA, April 17, 2003.

68. "The World of Harry Truman." *Time*, January 8, 1973. http://www.time.com/time/ magazine/article/0,9171,910501-4,00.html, 4.

69. Paul M. Edwards, *The Korean War: A Historical Dictionary* (Lanham, MD: Scarecrow Press, 2010), 204.

70. Several scholars have also read gun-slinging sheriffs and self-appointed vigilantes as fantasy substitutes for the president: Lawrence and Jewett specifically address this with regard to *The Lone Ranger*; Lawrence and Jewett, 2003. Film scholar John Lenihan applies a similar reading to Ford's 1950 film *Rio Grande*, which—coincidentally or not—was released five months after MacArthur was appointed UN commander. John H. Lenihan, *Showdown: Confronting Modern America in the Western Film* (Champaign: University of Illinois Press, 1980), 29–31. Stephen J. Mexal states that the villains in Delmer Daves's 1957 version of *3:10 to Yuma* clearly allude to the "communist threat." Mexal, "Locke, Liberalism, and Western Masculinity in *3:10 to Yuma*." In Jennifer McMahon and Steve Csaki, eds. *The Philosophy of the Western* (Lexington: University Press of Kentucky, 2010), 76. Stanley Corkin employs a similar reading of *High Noon*, which was released three months prior to the 1952 presidential election, in which Truman lost to Dwight D. Eisenhower. Corkin, *Cowboys as Cold Warriors* (Philadelphia: Temple University Press, 2004), 156–58. French remarks that this is a fairly common interpretation: European scholars Gunnar Oldin and Harry Schein, in separate papers, had made the same analogy in the mid-1950s; Philip French, *Westerns* (New York: Viking Press, 1973), 35. Daniel C. Hallin too allegorizes the president as a western marshal. Hallin, "We Keep America on Top of the World." In Todd Gitlin, ed. *Watching Television: A Pantheon Guide to Popular Culture* (New York: Pantheon, 1986), 12.

71. Bickel, 1971, 120.

72. Horsman, 1981, 247.

73. Churchill, 1998, 167–224.

74. Hage, 1998, 161.

75. Perhaps there may be some truth in this suggestion: There have been areas throughout history riven with chronic partisan and ethnic violence, which only the imposition of imperial rule has managed to quell. For example, the Balkans were relatively peaceful while the Ottoman Turks held a monopoly on violence there. Some minorities, such as Tamils in Sri Lanka, Jews in Tunisia, and Chinese in Indonesia, preferred living under British, French, and Dutch colonial rule. See Thomas Sowell, *Race and Culture: A World View* (New York: Basic Books, 1994), 124–25.

76. Corkin, 2004, 20.

77. Brenda M. Boyle, "Rescuing Masculinity." *Journal of American Culture* 34(2), June 2011, 152–53.

78. Ibid.

79. Robert Burgoyne, *Film Nation: Hollywood Looks at U.S. History* (Sacramento: University of California Press, 1997), 58; also see Shohat, 1993, 62.

80. James Marples, "The Lone Ranger, Freemasonry, and Texas-Ranger Ethics." Alvin Benemerito's Freemasonry Site. Accessed November 3, 2006. http:// www.masonic.benemerito.net/msricf/papers/>marples/marples-lone_ranger_freemasonry.pdf. On Trendle's membership see George W. Trendle, letter to Ines Wallace. December 12, 1956. George Trendle papers, Burton Collection. Detroit Public Library. Harry Truman and John Wayne were also members. "Famous Masons." Scottish Rite, Valley of San Diego. May 2009. http://www.scottishritesandiego.org/famousmasons.html.

81. William C. Smith, "God's Plan in America." *New Age Magazine* 58(9), September 1950, 551. Quoted in Phillip D. Collins and Paul David Collins, *The Ascendancy of the Scientific Dictatorship: An Examination of Epistemic Autocracy from the Nineteenth to the Twenty-first Century* (Bloomington, IN: iUniverse, 2004), 73. My purpose here is not to engage in a "witch hunt" against Freemasonry but merely to note, as the article demonstrates, that Scottish Rite at this time was an openly racist organization.

82. Ernest Lee Tuveson, *Redeemer Nation: The Idea of America's Millennial Role* (Chicago: University of Chicago Press, 1968), 126–28, 137–44, 150–56.

83. Horsman, 1981, 247.

84. Zackel, 2007.

85. Dorfman, 1983, 80.

86. Ralph Friar and Natasha Friar, *The Only Good Indian: The Hollywood Gospel* (New York: Drama Book Specialists, 1972), 189.

87. Vine Deloria, 1988, 202.

88. Moore, 1998, 145. Silverheels (real name Howard J. Smith), an actual Mohawk, deserves credit for bringing as much dignity as possible to this hackneyed role.

89. Joe Orlando and Edward Nelson Bridwell, "TV Scenes We'd Like to See." *Mad* 38, March 1958, 42.

90. Oscar Brown Jr., "The Lone Ranger." *Fresh* (Atlantic Records, SD 18106, 1974).

91. Doug Linder, "Lenny Bruce." University of Missouri, Kansas City, Faculty Projects, 2003. http://law2.umkc.edu/faculty/projects/ftrials/bruce/brucebiography.html. In 1968 John Magnuson made a short animated film, *Thank You Mask Man*, from one of Bruce's comedy routines. Lenny Bruce and John Magnuson, directors. *Thank You, Mask Man* (Columbus Productions, 1968). Posted on YouTube. http://www.youtube.com/watch?v=NZMTPGc0Z6A. Bruce died in 1966.

92. Bill Cosby, "The Lone Ranger." *I Started Out as a Child* (Warner Bros. Records, WS 1567, 1964).

93. *Tonight: Four Decades of "The Tonight Show" Starring Johnny Carson* (Respond 2 Entertainment, 2010). DVD.

94. Lyle Lovett, "If I Had a Boat." *Pontiac* (Curb Records, 471795, 1987).

95. Defoe, 2007, 270.

96. Rothel, 1981, 56, 191.

97. Vine Deloria, 1988, 202.

Chapter Three

Hollywood-ized History

Cochise in Broken Arrow *(1956–1958)*

> *One of the things the western is always about is America rewriting and reinterpreting its own past, however honestly or dishonestly it may be done.*— Philip French[1]

> *If there's a new stereotype to be found in prime time, it's that of the compassionate white government who is eager to make things right for American Indians.*— Christopher Campbell and Lori Edmo Suppah[2]

The fable of the winning of the West and its derivative dime-novel and western-movie genres served as a self-portrait of Euro-Americans, their relentless westward march, and their God-given destiny to dominate the continent. Peter Bayers calls it the dominant culture's aggrandizing image of itself: "These mythic stories, while suggestive of the *subconscious guilt* of whites, obscure the reality of the genocidal violence of this conquest." [3]

This chapter explores the development of the Regulator, in this instance represented by legendary Chokonen Apache leader Cochise. Much like Cheyenne chief Brave Eagle (*Brave Eagle*, 1955), Cochise is viewed as a Good Indian because he collaborates with the invaders in the march of Manifest Destiny. Cochise and Bureau of Indian Affairs agent Tom Jeffords act as a team: Jeffords handles smoothing over confrontations with avaricious white settlers and overzealous army officers, while Cochise keeps his Apaches in line, making sure they do not violate his treaty with the U.S. government. Although the two characters are (at first) presented as equals, it soon becomes apparent that Jeffords wields the real power in this relationship. Thus Tonto has gone from flunky to junior partner, and in later chapters we shall see how this developmental trajectory continues with the Indian enforcer taking on many of the trappings normally reserved for the white savior. We also see that the white savior is becoming more and more Indianized. Eventually these two figures, the Anglicized Indian and the Indianized white man, merge completely in Cordell Walker (*Walker, Texas Ranger*; discussed in chapter 8). This chapter also investigates *Broken Arrow*'s influence on the impact of the myth of the winning of the West during the 1950s as well its domestic political influence in its contemporaneous setting. Conceptualized as a liberal, pro-Indian text, *Broken Arrow* served to satisfy Euro-Americans' hunger for a sanitized history; at the same time, it obscured an issue that was crucial to American Indians during the Cold War years: termination.

A side issue but nonetheless important is how eager certain Hollywood executives were to incorporate nationalist messages into their products during the Cold War and how they volunteered to serve as "adjuncts of government," lending some credence to Althusser's claim that the media serve as Ideological State Apparatuses.[4] There are precedents for this. Prior to World War II, the Franklin D. Roosevelt administration made extensive use of the "carrot-and-stick" approach in enlisting Hollywood executives.[5] During the war, of course, Hollywood cooperated vigorously with the government, and Roosevelt's successor, Harry Truman, expected such cooperation to continue when the Cold War began to turn hot in Korea.[6] Executives at Twentieth Century-Fox (TCF) were particularly eager to serve as government propagandists. As Aleiss notes, TCF's 1950 film *Broken Arrow* "echoed government attitudes."[7] Even before Eisenhower's 1952 election, TCF executives were meeting with his inner circle, volunteering to insert nationalistic "propaganda" (actual term used by Darryl F. Zanuck) into TCF's films and television programs.

BACKGROUND OF THE SERIES

Prior to 1950, the American Indian in film had generally been portrayed either as an evil, murderous enemy or as an innocent child of nature—often both simultaneously—rarely landing between the two extremes.[8] *Broken Arrow* represented a notable effort to break this mold, as did *Brave Eagle* (1955). Although *Broken Arrow* favors the Noble Savage trope, with Chiracahua leader Cochise and his confederation consistently (and unrealistically) ennobled, it did make an attempt to present Apache characters in a reasonable manner. Apache characters are often portrayed as individuals instead of as an undifferentiated mass. More importantly, their attitudes are not monolithic. They are neither implacably hostile nor submissive. They behave as any group of men (there are few women) might when threatened with encroachment. Sometimes, like "regular" people, they are simply confused.

Cochise is pragmatic, rarely hotheaded or emotional. He takes time to mull over his decisions and operates according to the Apaches' unwritten laws and what he deems their best interests.

Comparisons between ABC's television version of *Broken Arrow* and *The Lone Ranger*, entering its seventh and final season in 1956 (also on ABC) illustrate how the western genre had changed in only a few years. *The Lone Ranger*'s early time slot (7:30 p.m. Thursdays), indicates it was meant for family viewing, while *Broken Arrow*'s later slot (9 p.m. Tuesdays) indicates it was aimed at adults. *Broken Arrow* was far more ambivalent than the melodramatic *The Lone Ranger*. Although *The Lone Ranger* purported to be (loosely) based on history, *Broken Arrow* insisted it was a completely accurate depiction of historical events. *Broken Arrow* was part of a trend toward adult-oriented westerns, which began with CBS's *Gunsmoke*.[9] Adult westerns were well-suited to the mind-set of the Cold War era: they promoted authoritarian attitudes and served to reassure a public beset by Cold War anxieties. Media historian J. Fred MacDonald writes,

> The [adult] western succeeded in part because it was a political morality play for the frightened, confused, and dispirited. . . . [TV westerns] affirmed the desirability and superiority of innocent, American-style civilization. . . . They justified nationalism in terms of a moral code that may have sounded secular but had at its base a [Puritan] religious heritage.[10]

Adult westerns allowed for more complex, ambiguous themes. For example, *Broken Arrow* acknowledged that the United States military committed atrocities against Apaches.

It is crucial to examine *Broken Arrow* in its Cold War context as well as developments that occurred in the western genre. Although Hollywood had exhibited intermittent cycles of pro-Indian films,[11] canonical westerns of the late 1930s and 1940s generally depicted Indians as homicidal maniacs with white settlers as innocent victims, a classic example being Ford's *Stagecoach* (1939). This approach, however, had to be moderated during the run-up to the war. The FDR administration enlisted Hollywood executives and producers in creating "a mythic multicultural America in order to persuade a somewhat reluctant populace to unite in the war against the Axis."[12] The federal government also needed to recruit minorities for military service, including American Indians, who enlisted in large numbers. Degrading representations of Indians as murderous, irrational villains therefore needed to be expunged from motion pictures. "Scheming businessmen and crooked bankers began to replace Indians as the frontier's villains."[13] This approach would need to be maintained during the Cold War in the face of Soviet critiques of U.S. society. There were still occasional throwbacks, such as *Arrowhead* (Charles Marquis Warren, 1953); however, "relatively few westerns [took] an overtly hostile attitude toward the Indians as *Arrowhead* [did]."[14] Television westerns, on the other hand, had been intrinsically revisionist from the beginning, because nationwide network coverage did not emerge until 1951. Hence there were few westerns on network television in which Indians were portrayed as murderous villains.[15]

BROKEN ARROW: THE 1950 FILM

All three versions of *Broken Arrow* were adapted from a "historical novel" by Elliott Arnold titled *Blood Brother* (1947), set in 1870s Arizona Territory, a region wrested from Mexico in 1848. TCF adapted the novel as a feature film in 1950, directed by Delmer Daves. It featured James Stewart as Tom Jeffords, a former U.S. Army scout who is "plumb sick and tired of killin.'"[16] Jeffords has left the army and is on his way to his new job as superintendent of a mail line in Tucson. His task is to get the mail through Apache territory. *Broken Arrow* was quite successful, winning two Academy Awards. It was widely considered a breakthrough in terms of progressive depictions of American Indians. There had of course been pro-Indian movies, but few harsh challenges to the myth of the winning of the West had been presented (although these would become acceptable in the 1970s). However, there were limits to this pro-Indian discourse. Edward Buscombe explains,

> The liberal position of the film [*Broken Arrow*] depends on viewing at least some of the Apaches as reasonable human beings, but this extends only to those who accept the process of assimilation, which was the policy of the US government at the time in which the film was set and for much of the twentieth century.[17]

This echoes Gray's *assimilationist* discursive strategy (discussed in more detail in my introduction and in chapter 4 on *Law of the Plainsman*).[18]

Broken Arrow offered as much of a critique of the myth of the Old West as could have been tolerated during the Cold War. The main challenge was to the myth that Anglo-Americans never started wars and never fought dirty. *Broken Arrow* shatters that myth. Some of its implications—critical of overzealous military officers; unenlightened "redneck" settlers; benighted federal officials; and genocidal policies—are in fact startling. It is an entirely new approach to relations with American Indians, one that downplays but does not overlook America's ugly moments in its battles with Apaches. This reflects a strategy that public-relations specialists now call "damage control," suggested to the state department in 1945 by communications consultant Harold Lasswell. The idea, Lasswell explained, was to "play up" the progress being made in race relations and avoid addressing anything negative.[19] This is exactly what *Broken Arrow* did. Thanks largely to the Supreme Court's

1954 ruling in the case of *Brown v. Board of Education of Topeka Kansas*, assimilation officially became the order of the day in government policy—and Hollywood's output followed suit.

THE 1956 TELEVISION PILOT: *TWENTIETH CENTURY-FOX HOUR*

A condensed adaptation of the film aired in May 1956 as episode 15 of *The Twentieth Century-Fox Hour*, an anthology series on CBS.[20] This episode served as a pilot for a weekly series, which appeared on ABC in September. Scenes from Daves's film were inserted into the *Twentieth Century-Fox Hour* episode, including part of the opening sequence, which begins with an extreme long shot of the open land, a generic convention identifying the western genre. This gaze at the land is more than just an appreciation for its natural beauty, however; it is more of a fetish (this "proprietary gaze" was discussed in detail in chapter 2 on *The Lone Ranger*). The land is, as usual, notably empty: unused, uncultivated, and therefore available for the taking.

That the land itself is the crucial issue is acknowledged in the first sentence of the episode's narration: Jeffords (played by John Lupton) intones, "This is a story of a land." A lone figure in the distance rides slowly toward the camera (figure 3.1). "These figures in this landscape are intrinsic to the appeal of the western."[21] He is visually connected to the mountains in the background; this association lends him strength.[22] The sky, which comprises about three quarters of the screen, however, dwarfs the character, but as we become more involved in his tale, he begins to loom larger as the viewer's identification with him grows (this effect, called *secondary* identification, is discussed in detail in chapter 9). The emphasis on the sky suggests that we are about to receive a message from the heavens, a message the rider carries in his mailbag; like the Lone Ranger, he is associated with the heavens. He emerges from behind a tall cactus that resembles a crucifix: a Christ-like figure, sent by God. The rider wears a white hat; in addition, his horse is mostly white. The horse's dark spots visually suggest that the rider, like the horse, is not *entirely* white. Unbeknownst to even himself, the rider will become, spiritually at least, Indianized. A jump cut shifts to a full shot of the character taken from a low angle, his head framed against the sky with clouds as a sort of halo.

The rider stumbles across a young Chiricahua boy who has been shot. In a sequence reminiscent of *The Lone Ranger*, Jeffords nurses the boy back to health, cleaning his wounds, and earning his respect. Apache scouts, including Geronimo, find them.[23] Because the boy vouches for this white man's good intentions, they let him go: "You did not kill. We will not kill—this time." Back in Tucson, with newfound respect for the Apaches' honor, Jeffords decides to learn their language and culture, which he does in short

Figure 3.1. A lone figure on horseback approaches in the distance, traveling from right to left (east to west).

order, illustrating his extraordinary intelligence and initiative. Jeffords then goes back to the Apaches' stronghold, finds Chiricahua leader Cochise (Ricardo Montalban), wins his respect, and "goes native." Jeffords becomes an adopted member of the tribe, marrying the beautiful, young Sonseeahray (Rita Moreno).[24] *The Twentieth Century-Fox Hour* episode contains an almost word-for-word transcription from the 1950 film—in which Jeffords has a confrontation with Colonel Bernall, who has come to re-recruit Jeffords as an army scout, and Ben Slade, a local rancher. Bernall openly advocates genocide: "Wiping out the Apache is the important thing right now." Jeffords takes issue with this remark. Rancher Slade, however, agrees with Bernall: "The Apaches burned my ranch last month. My wife was in the house." A heated exchange ensues:

Slade: Why won't you scout for the colonel?

Jeffords: That's private business.

Slade: War ain't private—not when women and children are being murdered.

Jeffords: At Big Creek, *we* murdered women and children.

Slade: Why not! Cochise started it, and it's up to every man to—

Jeffords: Get your facts straight, Slade! Cochise didn't start this war. It was started by a lieutenant fresh out of the East. Flew a flag of truce, which Cochise honored. Under it hung Cochise's brother and five more.

This is the scene that exposes the United States' dirty tactics in its war with the Apaches and lays bare the myth that U.S. forces are always "the good guys." Here the text overturns the conventional notions of who is savage and who is civilized. In Jeffords's eyes, the real savages are "greenhorn" Easterners, unenlightened and unsympathetic to Apache ways, who prefer to get on with the killing in the most efficient if dishonorable manner. Indeed, white settlers like Slade are usually depicted in this series as greedy, land-grubbing rubes, while federal officials such as Jeffords and some of his army cohorts are depicted as enlightened and altruistic, here to protect the rights of minorities. This would have constituted an important message during the Cold War and the civil rights struggle, which in 1956 was reaching critical mass.

THE WEEKLY SERIES

The project was quickly picked up by ABC and adapted into a weekly series; novelist Arnold was retained as story editor.[25] ABC aired seventy-two episodes, from September 26, 1956, to September 23, 1958. John Lupton returns as Jeffords; however, Ricardo Montalban as Cochise is replaced by Michael Ansara, who is of Egyptian-Arabic descent. Ansara had by this point already been typecast as an American Indian.[26]

In the TV version of *Broken Arrow*, Ben Slade's name is changed to Ben Cartwright.[27] Cartwright and Jeffords engage in an even more heated disagreement than the one in the pilot, with Cartwright demanding outright extermination of the Apaches. Cartwright here becomes even more bellicose and racist:

Cartwright: Wipe 'em out—slaughter 'em like the animals they are!

Jeffords: [Some dialogue appears to be excised] At Big Creek, *we* murdered women and children.

Jeffords then describes the U.S. Army's treachery against Cochise and his band of Chiricahuas. His assistant, Milt, corroborates, "It was a cold-blooded Judas trick." Given the McCarthy mentality of the day, this is remarkable for network television. The mere suggestion that the U.S. military had committed atrocities ran a risk of being construed as unpatriotic or "un-American." Apparently, however, the political demands for the United States to present itself to the rest of the world as nonracist during the civil rights struggle outweighed the demands of superpatriots like McCarthy. As an adult western, *Broken Arrow* was uniquely positioned to address Cold War anxieties, international hostilities, and domestic racial prejudice—yet it had to deal with these issues gingerly in order not to alarm advertisers, who recoiled from controversy.

Unlike the feature film and the *Twentieth Century-Fox Hour* adaptation, which both focus on Jeffords from the start, the opening sequence of the ABC-TV program presents the Apache and the white man as equals, giving them both the same amount of screen time.[28] This in itself was groundbreaking compared to most westerns, particularly *The Lone Ranger*, in which Tonto was more or less the Ranger's manservant.

The opening sequence of the TV series, typically, emphasizes the sky. It intercuts shots of both characters, each approaching the other on horseback from opposite directions. However, Jeffords looms much larger in the frame (figure 3.2); he also approaches from the left, a subliminal indication of his dominance, which will indeed be demonstrated eventually: as the series progresses, Cochise's role diminishes, and there are episodes in which he does not appear at all.[29]

The two meet in the middle ground (figure 3.2) and shake hands Indian-style, grasping each other's forearms (figure 3.3). Both are framed against the sky with clouds for haloes. The Indian character is thus beginning to take on the trappings of white savior.

There are other Apaches framed against the sky in godlike poses, but this device is generally reserved for Jeffords or Cochise. Both Jeffords and Cochise are photographed from low angles, making them appear taller than ordinary mortals (Ansara is already quite tall at six feet, three inches). Backlighting and halo effect, however, are reserved for Jeffords. In some shots, it looks as if the light of heaven has found his head (figure 3.4).

NONWHITE PHYSICALITY AND ISSUES OF MASCULINITY

A significant aspect of both the pilot and the series is that the Cochise character spends much screen time displaying his physique, with Ansara being the bulkier of the two actors. According to Hal Erickson, Ansara was hairy and had to shave his torso before every shoot.[30] The issue of body hair,

Figure 3.2. The white hero, framed against the sky, greets an unseen person.

or lack of it, has interesting implications. For one, American Indians are assumed to not have much of it, so Ansara's shaven appearance could be an attempt at verisimilitude. But there is the question of bodily display. In the 1950s, the idea of a human body on display was considered feminine.[31] That such displays were reserved for nonwhites—with the exception of Tarzan, who, spiritually at least, was not entirely white—implies that dark-skinned characters are physical, while Europeans are cerebral and must rely on brain rather than brawn.[32] The Indian is strong while the white man is clever. According to Richard Dyer, the reason white men's physiques are rarely if ever displayed in colonial narratives is because "[nakedness] may undo the remorseless insistence on difference and concomitant power carried by clothes and grooming. The exposed, white-male body is liable to pose the legitimacy of white-male power: Why should people who look [so ordinary] have so much power?"[33]

Not only are westerns a celebration of white conquest, they are a celebration of masculinity.[34] Even though nonwhites here are feminized by displaying their bodies, they evince a certain nature-based form of masculinity—still somewhat feminized, since nature is thought of as female. "Non-white masculinities appear from a dominant, white point of view to be in touch with

Figure 3.3. The figures meet in the middle; emphasis appears to be on Cochise.

'essential,' 'natural' masculinity . . . with which modern whites of European origin believe themselves to be out of touch."[35] So white men, who both produced and consumed these films and television programs, allowed a grudging respect for native masculinity. Yet "[r]epresentations of non-white masculinity pose a threat to white masculine identity insofar as they highlight the inadequacies of the latter,"[36] and indeed the Cochise character (as played by both Montalban and Ansara) is an impressive physical specimen, far more muscular than Jeffords. A rather elaborate mechanism must therefore be constructed in order to downgrade the Apaches' innate masculinity. Natives, then, evince a complex, dialectical—compromised—version of masculinity, whereas whites by contrast evince a less ostentatious, plain-folks type of masculinity. Indeed, most of their masculine power is in their brains.

THE WHITE-MALE GAZE

The question of "the gaze" has been around for decades if not centuries. Who is the "spectator" of a painting, a photograph, a film, a television program presumed (or imagined) to be? In more modern terms we might ask, who is

Figure 3.4. Jeffords with halo effect, signifying godliness.

the "target audience" for these images? In Dutch portraiture of the seventeenth century, for example, the spectator is often presumed to be the person who commissioned the painting, reflecting the viewpoint of the putative buyer, almost always a prosperous businessman.[37] In 1975, Laura Mulvey identified the filmic gaze as presumptively male.[38] That this is the case could be noted in the simple fact that nearly all movie directors are males. According to Mulvey, this male gaze *constructs* the viewer as a (heterosexual) male, whether or not the actual viewer happens to be male; he or she is put in the position of seeing things from the point of view the director evinces (which conceivably *could* be female but most likely is not). The camera itself, then, is assumed to be male.

There is more, however. Just as women are "other-ized" in much of Euro-American culture, others are often, through similar processes, feminized. American Indian (as well as other "foreign") bodies are "to be looked at" just as women are. Thus it is reasonable to conclude that the gaze Mulvey speaks of is not just a male gaze but a *white*-male gaze, and in the case of the Dutch portraits, it was that of an *affluent* or successful, middle-class male. So it is a bourgeois-white-male gaze. Dyer demonstrates repeatedly how the spectator of European and American art has always been presumed to be white and how art itself—as well as photography—celebrates whiteness.[39] This constructing of the spectator as a middle-class white male involves two types of identification: first, identification with the camera, that is, what it sees and how it sees it; and second, identification with a male character, usually the protagonist of the story.[40]

MORE INDIAN THAN THE INDIANS THEMSELVES

As noted in the introduction, as much as the Puritan and Pilgrim settlers professed to loathe American Indians, at the same time many envied them and wanted to be like them. A century later, beginning with Cooper's character Natty Bumppo (also known as Leatherstocking or Hawkeye), there came a growing fascination in U.S. literature with the hybrid white/Indian character. Leatherstocking was based on the real-life Daniel Boone.[41] The Indianized white man conforms to a recognizable pattern: he is an outsider to society, a somewhat wild (i.e., uncontrollable) man often opposed to conventional civilization and therefore willing to operate outside the law. He follows the laws of nature. Even the Lone Ranger (see chapter 2) had some Indian qualities, and those he lacked he was able to pick up quickly from his companion, Tonto, while Tonto became, in Zackel's term, an "apprentice white man."[42]

Broken Arrow's Tom Jeffords is a highly Indianized white man, a hybrid who embodies the best of both worlds. By episode 2 ("Battle at Apache Pass"), he has been appointed "Indian agent" in charge of the reservation, which means he is the Chiricahua Apaches' official authority figure. Cochise steps aside to become a spiritual leader: their relations are reminiscent of the historical collaboration between pope and emperor. By episode 7 ("Hermano"), Jeffords becomes a veritable superhero: he can do absolutely anything he sets his mind to. Already having been a lawyer, army scout, Indian fighter, prospector, mail manager, honorary Apache, Cochise's "blood brother," husband to the most beautiful Chiricahua maiden, and federal Indian agent, Jeffords, in the manner of Lieutenant T. E. Lawrence, organizes and leads of a group of Apache scouts who volunteer as U.S. Army aides to help find and defeat Geronimo and his "renegade" Apaches.[43]

Broken Arrow is notably akin to *Lawrence of Arabia* (1962) in that it requires a white man to organize and protect intellectually inferior Others from rapacious whites—as well as from competing natives.[44] Another starkly similar example is Kevin Costner's 1990 film *Dances with Wolves*, which Ward Churchill sardonically dubbed "Lawrence of South Dakota."[45] Both are tales of white men who reach enlightenment, transcending racism. Both Tom Jeffords and Lieutenant John Dunbar (Costner's character) are former army officers who now live among Indians. Both hark back to a literary archetype Richard Slotkin dubs "the Man Who Knows Indians."[46] The Man Who Knows Indians studies native culture and becomes—spiritually at least—part savage, so that he may use the Indians' savage ways against them. He is Indianized for the purpose of thinking like an Indian, thereby "beating them at their own game." However, there is the imminent danger this character may forget whose side he is on and end up "going native."[47] Indeed, Jeffords does exactly this, as does Dunbar. This is also the theme of James Cameron's film *Avatar* (2009), in which a U.S. military officer rescues a group of extraterrestrial natives who are under siege by colonizing forces. *Avatar*'s story line is remarkably similar to *Dances with Wolves*; both borrow liberally from Daves's 1950 film version of *Broken Arrow*. In these scenarios, the white hero is not only a phenomenal white man; he can do anything a native can do and better. He is an important man in both cultures. Jeffords understands Apache language and culture. He can outfight or outwit any Apache (with the possible exception of Cochise). He has his pick of the prettiest Indian maiden. "He's better at being an Indian than they are. . . . There's nothing a white man's better at than being an Indian."[48]

COCHISE AS ENFORCER OF ANGLO-AMERICAN LAW AND ORDER

Though it begins by presenting the Indian and the white man as ostensible equals and by presenting Indians as regular people, the TV version of *Broken Arrow* soon falls back on the simplistic Good Indian/Bad Indian dichotomy. Cochise, like Tonto, is a Good Indian because he helps the white man in his quest to control the land. Like Petain or Quisling, Cochise becomes an enforcer of the invaders' laws, rationalizing that he is looking out for his people's best interests. He has witnessed the white man's awesome technology (cannons) in action and reasonably concludes that if there were to be a full-blown war, his people would certainly lose. Cochise decides not to gamble what is left of their lives. He agrees that they will remain on the reservation. He signs a treaty with General Oliver O. Howard and gives his oath to uphold the bargain. What is more, Cochise personally executes Bad Indians and renegades who violate the peace by attacking whites. If Tonto in *The Lone Ranger* was, in the Friars' and Deloria's term, a fink,[49] then Cochise is a *superfink*, with powers and strength bordering on superhuman.[50]

Geronimo, the leader of a splinter group, is a Bad Indian. These renegade Apaches neither recognize nor obey the white man's laws nor Cochise's peace. They neither recognize nor respect private property. They invade settlers' homesteads, steal their livestock, rape their women. Cochise decides to help the U.S. Army capture and punish Geronimo. When Jeffords organizes a squadron of Apache scouts for the army, Cochise invokes their sacred duty to preserve his honor. In episode 13 ("Apache Massacre"), one of Geronimo's men commits the ultimate crime, attacking a white woman.

A white settler is eating dinner with his wife and daughter, looking very much like a 1950's-style nuclear family. The key light is on the woman, accentuating her blond hair and white skin, giving her an ethereal glow. She dominates the frame. Her blond locks contrast with the dark wall in the back; her hair is a golden halo. She is the madonna. The family lives in a cabin on what was Chiracahua land. The father hears a horse whinnying and goes to the door to investigate. He is on the lower right, indicating his forthcoming demise. The light emanates from inside the house, which represents the warmth of the Euro-American family. All else is darkness. While standing on the dark porch posing nobly, he is cut down by an arrow to his chest. The light reaches out to him, but he cannot respond.

A faceless Apache, seen only from behind, skulks about, as if he can't decide whether to go into the house. Finally, the Apache steps over the dying man. The Indian looks around furtively, then enters the doorway to the house. As he slips into the house, the mother shrieks. She is about to experience the "fate worse than death," a metonym for rape and miscegenation,[51] which happened to be the premise of Ford's film *The Searchers*, released that same year (1956). Protecting innocent, pure white women has long been cited as a pretext for extreme violence against supposedly animalistic and insatiable nonwhites.[52] According to Dyer, the sanctity of the white woman's body was "both the symbol of white virtuousness and the last word in the claim that what made whites special as a race was their non-physical, spiritual, indeed ethereal qualities."[53] In revenge, a group of angry townsfolk burn down the nearest Apache village, shooting and killing Apache women and children while the men are away hunting. This village, however, consists of Good Indians who have had no truck with Geronimo and his band of renegades. Jeffords barely manages to avert a resumption of all-out war with the Apaches. He swears to an enraged, skeptical Cochise—until now Cochise has rarely if ever lost his temper—that justice will be served by the white man's jurisprudence: "This was a small group of men—of criminals. This had nothing to do with the feelings of white men [as a whole]." Cochise, as usual, accedes to Jeffords's judgment, however reluctantly. After several tense mishaps, the perpetrators are captured, tried, and hanged. Jeffords and his U.S. Army associates are here to protect the Apaches—just as President Dwight Eisenhower and a squadron of National Guard troops would be on hand to protect black students enrolling in Little Rock's Central High School eight months after this episode was broadcast.

INDIANS AS STAND-INS FOR AFRICAN AMERICANS

It is useful to consider this program's implications in the context of the civil rights movement, which was then reaching critical mass. *Broken Arrow* appeared about two years after the 1954 *Brown v. Board of Education* ruling, which ordered public school desegregation. What is more, a year before this episode aired, fourteen-year-old Emmett Till had been brutally beaten to death in Mississippi because a white woman said he had whistled at her. Bus rider Rosa Parks had made her historic stand in Montgomery, Alabama, about nine months before the series' debut. Yet controversy had rendered African Americans nearly invisible on network television.[54] Advertisers did not want to remind viewers about civil rights issues, much less address them. Racial prejudice was too strong a subject for television.[55] In order to address issues of prejudice, television writers created *parables*, casting other minorities instead of blacks as its victims. Thus *Broken Arrow* may have contained veiled commentary on racial prejudice using American Indians as metaphorical stand-ins for African Americans. Philip French writes, "*Broken Arrow* obviously could be viewed as a plea for racial tolerance."[56] The program works hard to distance the federal government and its now-enlightened officials from the racist violence and ignorance of uneducated settlers in the Old West, exactly as President Eisenhower would find it necessary to distance the federal government from Arkansas governor Orval Faubus's blatant racism in refusing to obey the Supreme Court's desegregation order.[57]

BROKEN ARROW AS PUBLIC RELATIONS

Despite its occasionally startling accusations and admissions of atrocities, *Broken Arrow* defuses and absorbs these by "demonstrating" that, even in the face of daunting difficulties, the system works as promised—and always has. This would have been an important message in 1956 for African Americans as well as white, liberal advocates of integration. One of the series' projects, then, was instilling public confidence in government. This is not surprising. Network executives, usually eager to mollify both Congress and the executive branch, had developed a tradition of valorizing government officials. This began on radio in the 1930s with the valorization of "G-men" (FBI agents) and continued in film and later on television during the Cold War years.[58] Jeffords, too, was a federal agent: a Bureau of Indian Affairs official. *Broken Arrow*'s main message—trust your government—countered (and literally replaced) that of *The Lone Ranger*, whose primary message was that the system *does not* work—otherwise why would it be necessary for the Ranger to take the law into his own hands by becoming a masked vigilante?

BROKEN ARROW AND TERMINATION

If the 1950 film version of *Broken Arrow* "echoed government attitudes," as Aleiss asserts,[59] this was even more true of the ABC-TV series, which worked hard to make the federal government appear as if it were protecting American Indians—when in reality it was allowing private interests to plunder native resources. Beginning in 1950, Congress empowered itself to terminate—at its own discretion—the 389 treaties it had long ago ratified with indigenous nations.[60] From 1954 to 1964, Congress selectively revoked the sovereign status of 109 nations. Some filed suit, but the Supreme Court upheld the government's position. Although Congress had proposed "termination" as a way to help balance the federal budget, the policy was also supported by so-called reformers as a way to force American Indians to assimilate.[61] More crucially, however, it served as a pretext for a land-and-resources grab that benefited mining interests. The Cold War, with its proliferation of nuclear weapons, had created a huge demand for uranium—much of which was found on Indian lands.

Under constant attack from Soviet (and French) newspapers, the federal government needed to cultivate a better image with regard to its treatment of American Indians. *Broken Arrow* supported the government's position, employing the "damage-control" strategy Lasswell had recommended to the state department in 1945. The essence of damage control (in public-relations terms) is to admit that there are or have been serious problems but the responsible entity is working hard to ameliorate them. The hope is that this will minimize public criticism. This is exactly what *Broken Arrow* does in its admission of army atrocities.

TCF executives eagerly cultivated a position as "adjunct of government." Even before Eisenhower's election, TCF's vice president of production, Darryl F. Zanuck, had been in contact with Eisenhower's top-level strategists, including Zanuck's close friend, Vice President Richard Nixon. Three months after Eisenhower took office, Zanuck personally met with members of Eisenhower's Psychological Strategy Board, volunteering to insert "propaganda" messages (his term) into TCF's films.[62]

FROM GENOCIDAL TO BENEVOLENT RACISM

Whether or not TCF's retrospective "reframing" the federal government as protector of American Indians was legitimate, stories like the ones in *Broken Arrow* simply shifted their ideological position from an obsolete, genocidal form of racism (personified by greedy white settlers) to a more benign but condescending form in which the natives become childlike, in need of protection by their white "fathers." Such texts depict natives as helpless objects in need of management.[63] This tactic allows white audiences to identify with the "good" white protagonist rather than his less evolved predecessors who clamored to "wipe out" Indians. Thus younger, liberal audience members could distance themselves from older, more blatant white supremacists, much the same way as Eisenhower would do with Faubus. Courtney W. Bailey notes that the way the media individualizes issues rather than showing them as systemic, society-wide problems allows mainstream culture

> to continue to understand itself as liberal and tolerant in the face of a crisis that might indicate otherwise. . . . This audience represented successful neoliberal citizens . . . *absolved from needing to consider collectivity and systemic inequalities*. . . . This freed this audience of the need to examine its own complicity in structures of privilege.[64]

In 1956, such a crisis was being unmasked on a near-daily basis by civil rights protestors and their adversaries. In order to avoid implicating viewers, it was crucial that the network blame the problem on ignorant individuals rather than on U.S. society as a whole. Thus *Broken Arrow* works to absolve its viewers of racism, flatters them as progressive, and differentiates them from the ignorant ones. It positions modern, liberal viewers on the correct side of history. Social critic and English professor Ed Guerrero adds, "This representational gambit has served dominant society well. . . . Most [people] in that vast consumer audience [can thus be] personally distanced from social inequality's systematic workings as they look . . . for *psychic comfort* and assurance that the society is making racial progress."[65] As Barthes predicted in 1957 (the year after *Broken Arrow* aired), the primary purpose of myth is to provide a plausible escape from a moral dilemma.[66] It is not a coincidence that a French intellectual would make such an observation and that it would apply to the United States: France, at precisely this time, was experiencing a similar crisis stemming from its expulsion from Vietnam and its (ultimately) vain struggle to retain Algeria as a colony.

In addition to myth, there are other methods used to elide painful truths and soothe ideological crises. In a 1987 article, Khatib/Clark explains a psychological process he calls *temporal displacement*, which he describes as a defense mechanism that deals with embarrassing, contradictory, or painful information.[67] This "retemporalization" compartmentalizes unwelcome information in another period; in other words, it is buried in the past. This offers an explanation as to why so many films and television series with American Indians are situated in the 1800s: it is a way of suggesting that crimes perpetrated against them are a thing of the past and that it is too late to do anything about them.[68] In addition, retemporalizing is a way of pushing Indians into a mythical realm, which in 1956 served to distract the public from actual problems that existed at the time, which mainstream viewers most likely preferred to avoid and would continue to haunt and embarrass U.S. leaders in the international arena—issues such as structural racism, poverty, and termination.[69] Damage control entailed acknowledging such problems, yet these became buried in a distant past. Peter Bayers writes, "Yes, genocide was committed on US soil, but it is easier to avoid this truth if Indians in films like *Dances with Wolves* are consigned to a mythical realm [where] they constitute no threat to the established order."[70]

WESTERNS AND HISTORICITY

Allusions to historical authenticity have always been part of the western genre's appeal.[71] Both Daves's 1950 feature film and *The Twentieth Century-Fox Hour* version of *Broken Arrow* purport to be based on actual fact. In the opening narration, Jeffords announces, "What I have to tell happened exactly the way you see it. The only change will be that when the Apaches speak, they will speak in our language. What took place is part of the history of Arizona." Arnold himself insisted that "the main events in [his novel *Blood Brother*] are entirely true."[72] In actuality, very little of the story is true, and the parts that are based on real events are distorted, manipulated, reframed, and retemporalized in order to make the federal government look better. This should not be surprising in view of Zanuck's arrangement with the Eisenhower administration and Hollywood's tradition of cooperating with the Department of State.

However true to history *Broken Arrow* purports to be, it omits many crucial facts and distorts many others. For one, Cochise died two years after signing the 1872 treaty, whereupon the government breached the treaty: the Bureau of Indian Affairs decided to consolidate the Chokonen (Cochise's group) with other Chiricahua groups in New Mexico.[73] This is exactly the situation Cochise had resisted. The government had framed the Chokonens' displacement in terms of teaching them agriculture, but the more salient fact is that massive lodes of copper ore had been discovered on Chokonen land.[74] Copper is essential for the electronics industry and in the 1870s was required for telegraph lines. What happened in 1874, then, recurred in the 1950s, a scenario *Broken Arrow* served to obfuscate: The government, far from protecting Indians' interests, furthered the interests of corporate mining operations. History would repeat itself in the 1950s in the form of termination, but this time the resource would be uranium (for nuclear weapons). Thus *Broken Arrow* was an Orwellian history that absolved anxious Euro-American viewers of guilt and allowed them to continue to view their elected officials as protecting American Indians while they were actually displacing them and exploiting their resources—and poisoning them with radioactive waste.[75] Film scholar Douglas Pye explains that one of the cultural functions of the western is to *obscure* rather than to elucidate history: "These fictions exist in a sense outside time, creating a world that announces itself as the past, even 'history,' yet operates ideologically to suppress knowledge of both the past and the present."[76] Stanley Corkin agrees: most westerns can be approached as "fictional texts that strategically rewrite documentary evidence."[77] As Russell Means points out, this process allows Euro-Americans to remain ignorant of their own history.[78]

NOTES

1. Philip French, *Westerns* (New York: Viking Press, 1973), 24.
2. Christopher Campbell and Lori Edmo Suppah, "Television's Troubling Indian Images." *Television Quarterly* 33(4), Spring 2003, 25.
3. Peter L. Bayers, "The US Mint, the Lewis and Clark Bicentennial, and the Perpetuation of the Frontier Myth." *Journal of Popular Culture* 44(1), February 2011, 42, 48.
4. Althusser seemed to imply that this was their primary or exclusive mission. However, in a for-profit system, making money remains the primary motivation. If there is an underlying ideology, it would be that of glorifying consumption itself. Manufacturing propaganda to suit the government's needs *might* be included in a list of priorities—since broadcast television is regulated and thus operates at the mercy of government. Nonetheless, there have been occasions in which the media openly challenged the government; however, these are extremely rare.
5. John E. Moser, "Gigantic Engines of Propaganda: The 1941 Senate Investigation of Hollywood." *Historian* 63(4), Summer 2001, 73.
6. Truman issued Presidential Proclamation 2914 on December 16, 1950, which allowed his administration broad powers of censorship. Anthony Leviero, "President Declares a National Emergency." *New York Times*, December 17, 1950, A1. Truman's state of national emergency, declared in 1950, remained in effect until 1972. Since the Korean War, the executive branch has claimed the power to shut down all broadcasting in the event of a national emergency; however, this power has never been utilized. "Radio Bill Hearing Set." *New York Times*, February 11, 1951, A40. Also see "Govt. Gives Plans of B'caster Role in Civil Defense." *Variety*, April 18, 1951, 4. In 1962, President Kennedy authorized complete seizure of all communications industries in Executive Order 10995. President Nixon affirmed and expanded these powers in 1969 with Executive Order 11490.
7. Angela Aleiss, "Hollywood Addresses Postwar Assimilation: Indian-White Attitudes in *Broken Arrow*." *American Indian Culture and Research Journal* 2(1), October 1987, 75.
8. Norman A. Scotch, "The Vanishing Villains of Television." *Phylon* 21(1), Fall–Winter, 1960, 58–62; Robert F. Berkhofer Jr., *The White Man's Indian: Images of the American Indian from Columbus to the Present* (New York: Alfred A. Knopf, 1978), 71. Raymond William Stedman, *Shadows of the Indian* (Norman: University of Oklahoma Press, 1982), 248–50. Pamela Wilson, "Confronting the "Indian Problem": Media Discourses of Race, Ethnicity, Nation, and Empire in 1950s America." In Sasha Torres, ed. *Living Color: Race and Television in the United States* (Durham, NC: Duke University Press, 1998), 41.
9. We have seen (in chapter 2) how CBS chairman William Paley was responsible for the development of *Gunsmoke* (1955). Future western-movie director Sam Peckinpah wrote ten episodes of *Gunsmoke* as well as three of *Broken Arrow*.
10. J. Fred MacDonald, *Television and the Red Menace: The Video Road to Vietnam* (Westport, CT: Praeger, 1985), 139–40. The religious heritage of westerns has also been noted by Peter Homans, "Puritanism Revisited: An Analysis of the Contemporary Screen-Image Western." In Jack Nachbar, ed. *Focus on the Western* (Englewood Cliffs, NJ: Prentice-Hall, 1974), 84–92; and Michael T. Marsden, "Savior in the Saddle: The Sagebrush Testament." In Jack Nachbar, ed. *Focus on the Western* (Englewood Cliffs, NJ: Prentice-Hall, 1974), 93–100.
11. Aleiss, 1987, 67.

12. Donald Hoffman, "Whose Home on the Range? Finding Room for Native Americans, African Americans, and Latino Americans in the Revisionist Western." *MELUS* 22(2), Summer 1997, 45.

13. Angela Aleiss, *Making the White Man's Indian: Native Americans and Hollywood Movies* (New York: Praeger Publishers, 2005), 70.

14. Philip French, "The Indian in the Western Movie." In Gretchen M. Bataille and Charles L. P. Silet, eds. *The Pretend Indians: Images of Native Americans in the Movies* (Ames: Iowa State University Press, 1980), 104.

15. However, viewers could see hundreds of canonical westerns in the form of reruns of old B movies, which appeared in syndication.

16. The film rights to the novel were purchased in 1948 by Burt Lancaster and Harold Hecht (d.b.a. Norma Productions) and sold to MCA in 1949. MCA in turn sold the rights to TCF with the stipulation that its client James Stewart be "attached" as star.

17. Edward Buscombe, *100 Westerns: BFI Screen Guides* (London: British Film Institute, 2006), 17.

18. Also see Herman Gray, "The Politics of Representation in Network Television." In Horace Newcomb, ed. *Television: The Critical View*, 6th ed. (Oxford: Oxford University Press, 2000), 294–95.

19. Harold D. Lasswell, "Information Policies Concerning Russia (Rough Draft)." Memo to William Benton. October 22, 1945. National Archives RG59, Assistant Secretary for Public Affairs, Box 13, Correspondence, 1945–48.

20. *The Twentieth Century-Fox Hour.* Episode 15: "Broken Arrow" (Twentieth Century-Fox Pictures/CBS-Television, May 2, 1956).

21. Richard Dyer, *White* (London: Routledge, 1997), 34.

22. Will Wright, *Six Guns and Society* (Berkeley: University of California Press, 1977), 58, 82.

23. In the 1950 film version of *Broken Arrow*, Geronimo was played by Jay Silverheels, whose powerful performance went uncredited because he was under exclusive contract to the producers of *The Lone Ranger*. Silverheels reprised his role as Geronimo in the 1956 western film *Walk the Proud Land.*

24. American Indian characters are often played by Latinos and Latinas (Montalban is from Mexico; Moreno from Puerto Rico). This is not surprising, as many Mexicans and other Latinos have mestizo (mixed Spanish and Native American) heritage and are often dark skinned.

25. "Full Cast and Crew for Broken Arrow." Internet Movie Database. Accessed April 2, 2009. http://www.imdb.com/title/tt0048848/fullcredits#cast.

26. In 1952 Ansara played Shawnee chief Tecumseh's brother, Tenskwatawa the Prophet, in the film *Brave Warrior* (Jay Silverheels played Tecumseh). He also appeared as Angry Horse in the 1956 film version of *The Lone Ranger*. Ansara's image as an Apache was refined in 1954 when he played Apache Joe in the movie *Three Young Texans.* "Michael Ansara—Perennial Indian." *TV Guide*, April 1960, 29.

27. Ben Cartwright was also the name of the father figure played by Lorne Green on the western series *Bonanza*, which debuted in 1959.

28. A few other details were changed as well. For some reason—perhaps at the urging of Apache advisers—the ceremony of the White Painted Lady was renamed the "Dance of the Sunrise ceremony with the young girl."

29. Ansara may have been dissatisfied with the role. He told a *TV Guide* interviewer that "the acting range was rather limited. Cochise could do one of two things: stand with his arms folded looking noble or stand with his arms at his sides looking noble." "Michael Ansara—Perennial Indian," 1960, 29.

30. Hal Erickson, "Michael Ansara: Biography." All Movie Guide. Accessed February 28, 2009. http://www.allmovie.com/cg/avg.dll.

31. Dyer, 1997, 146. Martin Pumphrey, "Why Do Cowboys Wear Hats in the Bath? Style Politics for the Older Man." *Critical Quarterly* 31, October 1989, 78–100. Reprinted in Douglas Pye and Ian Cameron, eds. *The Movie Book of the Western* (London: Cassel, 1996), 52–55.

32. The films of Rudolph Valentino in the 1920s were also an exception. Graeme Turner, *Film as Social Practice* (London: Routledge, 2000), 139.

33. Dyer, 1997, 146.

34. Lutz P. Koepnick, "Siegfried Rides Again: Westerns, Technology, and the Third Reich." *Cultural Studies* 11(3), 1997, 418; Buscombe 2006, x; Dyer, 1997, 35; Pumphrey, 1996, 51; Wright, 1977, 58, 82.

35. Ramsay Burt, *The Male Dancer: Bodies, Spectacle and Sexuality* (London: Routledge, 1995), 117.

36. Ibid.

37. Wayne Franits, *Dutch Seventeenth-Century Genre Painting: Its Stylistic and Thematic Evolution* (New Haven, CT: Yale University Press, 2004), 12.

38. Laura Mulvey, "Visual Pleasure and Narrative Cinema." *Screen* 16(3), Autumn 1975, 11.

39. Dyer, 1997.

40. We have already seen these identificatory phenomena at work, in the opening sequence of the early *Lone Ranger* episodes, in which the Ranger scans the horizon with his proprietary gaze (see figure 2.3 in chapter 2).

41. Frederick Zackel, "Robinson Crusoe and the Ethnic Sidekick." *Bright Lights Film Journal* 58, November 2007. http://www.brightlightsfilm.com/30/crusoe1.html.

42. Ibid.

43. General George Crook called the Chiracahuas "the most subordinate" of his scouts. George Crook, *Resumé of Operations against Apache Indians, 1882 to 1886* (London: Johnson-Taunton Military Press, 1971), 364. U.S. forces aiso recruited scouts from the Yuma, Mohave, Mescalero, and Tonto bands.

44. Although the film version of *Broken Arrow* (1950) preceded David Lean's *Lawrence of Arabia* (1962), Lawrence's memoirs had been in publication since 1922. A film had been in the works since 1934 but for various reasons—including interference from the British government—production was stalled several times. See James Chapman and Nicholas J. Cull. *Projecting Empire: Imperialism and Popular Cinema* (London: I. B. Tauris, 2009), 10, 93.

45. Ward Churchill, *Fantasies of the Master Race: Literature, Cinema, and the Colonization of American Indians* (San Francisco: City Lights Books, 1998; originally published 1992), 239.

46. Richard Slotkin, *The Fatal Environment: The Myth of the Frontier in the Age of Industrialization, 1800–1890* (Norman: University of Oklahoma Press, 1998), 96.

47. An interesting real-life example of this is the story of Thomas Morton, an Englishman who helped establish the colony of Merrymount, in what is now Quincy, MA. Morton, a royalist who disliked Puritans, was sympathetic to the Pequots. He "went native" in 1627, infuriating his Puritan neighbors. He was captured, put in stocks, and exiled to the Isle of Shoals (in nearby New Hampshire). Morton escaped by ship to England, where he sued the Massachussetts Bay Co., the organization behind the American Puritans. With the support of King Charles, who also disliked Puritans, the company's charter was revoked. Morton became a celebrity in England, yet he returned to North America in 1642, where he continued to face persecution. He died in Maine in 1647 at age seventy-one.

48. Ward Churchill, in *Images of Indians: How Hollywood Stereotyped the Native American*, directed by Chris O'Brien and Jason Witmer (Starz Encore Entertainment, 2003).

49. Ralph Friar and Natasha Friar, *The Only Good Indian: The Hollywood Gospel* (New York: Drama Book Specialists, 1972), 189; Vine Deloria Jr., *Custer Died for Your Sins: An Indian Manifesto* (Norman: University of Oklahoma Press, 1988), 202.

50. I would surmise that Cochise's characterization, as well as Jeffords's and others like him, may have been influenced by the comic-book hero Superman, who appeared on U.S. television during this period (1952–1958).

51. J. P. Telotte, "A Fate Worse Than Death." *Journal of Popular Film and Television* 26(3), 1998, 127.

52. Dyer, 1997, 25–27; Arthur M. Eckstein and Peter Lehman, *The Searchers: Essays and Reflections on John Ford's Classic Western* (Detroit: Wayne State University Press, 2004), 311. According to Richard Slotkin, certain groups of western Indians, unlike their eastern

counterparts, *did* practice ritualized rape and torture on their victims. Richard Slotkin, *Regeneration through Violence: The Mythology of the American Frontier, 1600–1860* (Norman, OK: University of Oklahoma Press, 1973), 357. I am not sure whether he includes Apaches in this category.

53. Dyer, 1997, 127.

54. Melvin P. Ely, *The Adventures of Amos 'n' Andy: A Social History of an American Phenomenon* (New York: Free Press, 1991), 238.

55. Kaufman in Pepe Arciga, *Los Angeles Mirror*, August 26, 1957. Reprinted in *Los Angeles Times*, August 27, 2007. http://latimesblogs.latimes.com/thedailymirror/2007/08/pepe-arciga-1.html#more.

56. French, 1973, 81.

57. Ironically, Faubus may have been more of a moderate on racial issues than Eisenhower. Roy Reed, *Faubus: The Life and Times of an American Prodigal* (Fayetteville: University of Arkansas Press, 1997), 355; also see Watson, 1994, 90. However, this being politics, it was the public perception (and posturing) that counted most.

58. Tricia Jenkins, *The CIA and Hollywood: How the Agency Shapes Film and Television* (Austin: University of Texas Press, 2012), 38. Also see Philip D. Smith and Kristin Natalier, *Understanding Criminal Justice: A Sociological Perspective* (Thousand Oaks, CA: Sage Publications, 2004), 161, and William Marling, "Detective Fiction on Radio, TV, and in Comics." Detnovel.com. http://www.detnovel.com/OtherMedia.html.

59. Aleiss, 1987, 75.

60. House Joint Resolution 490, August 11, 1950. Introduced by Representative Reva Beck Bosone, July 1950. See Warren Metcalf, *Termination's Legacy: The Discarded Indians of Utah* (Lincoln: University of Nebraska Press, 2007),75–76. Also see Drinnon, 1987, 170–79.

61. Larry J. Hasse, "Termination and Assimilation: Federal Indian Policy, 1943–1961." PhD thesis, Washington State University, 1974, 114.

62. Jackson Committee, "March Meeting Notes." April 10, 1953. Washburn Abbott Papers, Box 7. Dwight D. Eisenhower Library, Abilene, Kansas. Also see Darryl F. Zanuck, telegram to unknown recipient. Department of State correspondence, June 20, 1956. Pre-Presidential Papers of Richard M. Nixon. General Correspondence, Series 320. Richard M. Nixon Library, Yorba Linda, CA. Zanuck had been in charge of the 1950 movie version of *Broken Arrow*, but by the time the ABC series appeared in 1956 had resigned to become an independent producer under contract to TCF. Zanuck, however, remained TCF's largest shareholder. George F. Custen, *Twentieth Century's Fox: Darryl F. Zanuck and the Culture of Hollywood* (New York: Basic Books, 1997), 357. TCF's chairman, Spyros Skouras, a stalwart Republican party fundraiser, remained in contact with the White House. "May 1958 Staff Notes (2); Abbott Washburn on Spyros Skouras." DDE Diary Series. Box 32. Papers as President (Ann Whitman Files), 1958. Dwight D. Eisenhower Library, Abilene, Kansas.

63. Ghassan Hage, *White Nation: Fantasies of White Supremacy in a Multicultural Society* (Annandale, NSW: Pluto Press, 1998), 93.

64. Courtney W. Bailey, "Coming Out as Homophobic: Isaiah Washington and the *Grey's Anatomy* Scandal." *Communications and Critical/Cultural Studies* 8(1), March 2011, 6, 17; emphasis added. Herman Gray too revisits this theme often.

65. Ed Guerrero, "The Black Man on Our Screens and the Empty Space in Representation." *Callaloo* 18(2), 1995, 396, emphasis added.

66. Roland Barthes, *Mythologies* (London: Paladin, 1972), 129.

67. Syed M. Khatib (f.k.a. Cedric C. Clark), "The Reduction of Cognitive Conflict through Spatial and Temporal Displacement." *Journal of Intergroup Relations* 15(4), Winter 1987, 28–31.

68. Chon A. Noriega, "Birth of the Southwest: Social Protest, Tourism, and D. W. Griffith's *Ramona*." In Daniel Bernardi, ed. *The Birth of Whiteness: Race and the Emergence of US Cinema* (New Brunswick, NJ: Rutgers University Press, 1996), 219.

69. Churchill, 1998, 168–69.

70. Peter L. Bayers, "The US Mint, the Lewis and Clark Bicentennial, and the Perpetuation of the Frontier Myth." *Journal of Popular Culture* 44(1), February 2011, 43.

71. Aleiss, 2005, 10–11.

72. Elliott Arnold, *Blood Brother* (New York: Duell, Sloan & Pearce, 1947), x.

73. Unwilling to share limited territory with competing Apache groups, Geronimo's group renewed its rebellion, escaping to Mexico between raids. Jeffords was again recruited by the army in 1886 to help find and capture Geronimo. This was reflected in episode 7, when Jeffords leads a cadre of "Indian scouts" in the hunt for Geronimo. However, the program's chronology is off by about twelve years.

74. Capstone Press Geography Department, *Arizona: One Nation* (Mankato, MN: Capstone Press, 2002), 25–27, 33.

75. Ward Churchill, *Struggle for the Land: Native North American Resistance to Genocide, Ecocide, and Colonization* (San Francisco: City Lights Books, 2002), 246.

76. Douglas Pye, "Genre and History." In Douglas Pye and Ian Cameron, eds. *The Movie Book of the Western* (London: Cassel, 1996), 111.

77. Stanley Corkin, *Cowboys as Cold Warriors* (Philadelphia: Temple University Press, 2004), 37.

78. In *Images of Indians*, 2003.

Chapter Four

Son of Tonto

Sam Buckhart in Law of the Plainsman, *1959–1960*

> *Colonization . . . takes peoples' stories away and assigns them supportive roles in stories that empower others at their expense.*—Davidson Loehr[1]

The Lone Ranger had his sidekick, Tonto, a sort of "apprentice white man," helping him administer frontier justice. Two years later, a series called *Law of the Plainsman* emerged on NBC. *Plainsman*, which ran for only one season, was very much like an extension of *The Lone Ranger*, except this time the Indian got to ride the white horse—and had a badge to boot. *Plainsman* is remarkable because it was the first U.S. network series to feature an American Indian in the role of full-fledged lawman. In this sense the American Indian fulfills Clark's prediction that minority characters must pass through the stage he calls "Regulation," in which they become enforcers of the dominant group's norms, proving their loyalty to the dominant culture. In *Law of the Plainsman*, Apache warrior Buck Heart leaves his group, heads east to attend Harvard, and—after being sufficiently Europeanized—returns to the West to become a U.S. marshal, an enforcer of the dominant group's law and order.

Law of the Plainsman was developed by Jules V. Levy, Arthur Gardner, and Arnold Laven (d.b.a. Cardiff Productions) in partnership with Hollywood-based Four Star Productions, partly owned by actor Dick Powell. NBC aired thirty episodes beginning October 1, 1959.[2] *Plainsman*'s central character, Deputy Marshal Sam Buckhart, had been introduced in two episodes of Four Star's *The Rifleman*. Buckhart is played by Michael Ansara. Ansara, an Arab American born in Syria, is swarthy looking with black hair, and he wears black eyeliner to make him look even darker.[3] Ansara's image as an

Apache had already been solidified during two years as Chiracahua leader Cochise on the television version of *Broken Arrow* (ABC-TV, 1956–1958), so he brought a strong intertextual connection to the series.

Plainsman, however, was more a conceptual extension of *The Lone Ranger* than of *Broken Arrow*. *Plainsman* and *Lone Ranger* shared a similar "backstory": just as a young Tonto had saved the future Texas Ranger, Buck Heart, a young Indian boy, tries to save a U.S. Army officer's life. The officer dies, but he leaves Buck Heart a large sum of money with the proviso that he use it to get an Anglo-American education—in other words, to become fully assimilated into Anglo-American society. Buckhart's religion is never mentioned—is it Apache religion or Christianity or some combination of the two?—but he occasionally spouts nebulous Christian platitudes. After earning a law degree, Buckhart returns to Santa Fe to become a deputy of federal marshal Andrew L. Morrison. He soon becomes Morrison's most valued employee. *Plainsman* is closely related to *The Lone Ranger* in other aspects. For one thing, it appeared at 7:30 p.m. on Thursdays, which had been *The Lone Ranger*'s time slot at ABC for eight years (1949–1957). Indeed, *Plainsman* might be thought of as a sequel of sorts to *The Lone Ranger*—along the lines of, say, *Son of Tonto*, in which the Indian sidekick gets a promotion, along with the requisite white horse. *Plainsman* seems to be based on the premise: What might happen were the Ranger to disappear? Would Tonto carry on the fight for law and order? What if the Ranger and Tonto were combined into one character?

Both the Ranger and Buckhart are based on biblical avenging-angel figures derived from the Old Testament. This chapter will examine some of the visual tropes that signify Buckhart's divine authority, indeed his divinity. *The Lone Ranger* and *Plainsman* differ in some crucial ways, however. Most importantly, *The Lone Ranger* comprised a right-wing text, while *Law of the Plainsman* was centrist/pluralist. Peter Biskind explains that texts that feature vigilante figures tend to convey extremist right-wing sentiments because they value individual initiative above consensus—even above the law—implying that the legal system is not only dysfunctional but hopeless. This implies that the system is beyond reform and should be scrapped—or at best ignored. The centrist/pluralist model, on the other hand, urges citizens to work within the system and allow the "experts" to do the jobs for which they were appointed.[4] Whereas the Lone Ranger operates entirely outside the system and society, Buckhart is a duly appointed officer of the law as well as a fully integrated—though often undervalued—member of the community. Ultimately, *Law of the Plainsman* endorses the federal government and its law-enforcement officers. In centrist texts, "we will have seen justice served [by the system] and know that in the United States, government is indeed good."[5] However, *The Lone Ranger*'s subtext implied the opposite: government is useless if not corrupt; thus it is up to the righteous individual to ignore or override it by taking the law into his own hands. The Lone Ranger–Tonto link comprises a typical "buddy" pairing.[6] Thus, ironically, the Ranger was not truly a "lone" ranger, since he almost always had Tonto with him. Lawman Sam Buckhart, on the other hand, is a loner.

Both *The Lone Ranger* and *Law of the Plainsman* must be situated in their historical contexts and their responses to the ideological needs of the Cold War as well as the civil rights movement, which were inextricably intertwined. It is fairly well established that Cold War television in the United States often contained an underlying nationalist function.[7] The ideological advantages of endorsing a minority figure as federal official in this context are clear: this symbolic uplifting of a "colored" (i.e., nonwhite) man into a position of authority would have delivered an important subtext to the viewing public, supporting the federal government's position during the Cold War *and* during the civil rights movement, both of which were in full swing in 1959. The message: "Be patient—things are changing: the colored man is moving up in the world." Not only did Americans at home require this message, but it served as a rebuttal of sorts to the Soviet (and French) newspapers that vilified the United States for supporting its suppression of minorities.[8]

Buckhart also resembles *Superman*'s Clark Kent (as does Cordell Walker in *Walker, Texas Ranger*): he has a secret identity as an at-will Indian (Buckhart as an Apache; Walker as a Cherokee). Buckhart also exemplifies the ways in which the Anglo-American (cerebral) figure, personified by the Lone Ranger, and the Native American (physical) figure, personified by Tonto, begin to merge into one character representing "the best of both worlds."[9] Another consideration is whether American Indians are being used in this series as metaphorical "stand-ins" or surrogates for all U.S. minorities—African Americans in particular—and how the series supported the federal government's position during the civil rights struggle. On the other hand, a contradictory argument could be made that American Indians were being used here as a "model minority," a conservative trope that has been used to *criticize* African Americans.

BACKGROUND OF THE SERIES

Although Buckhart is clearly an extension of Tonto, the differences between Tonto and Buckhart are notable. As already noted, Tonto operates outside the law, assisting the Ranger in his illegal vigilante efforts, while Buckhart is a duly appointed officer of the federal government. In addition, Buckhart is highly educated: he speaks exemplary English, has a Harvard law degree, and reads Socrates in the original Greek (episode 22, "A Question of Courage").[10] Buckhart, who has internalized the Anglo-American concept of social mobility—certainly not an Apache value—has been raised to a higher social status by the dead officer's money, which purchased his elite education. Buckhart clearly relishes this upward mobility, which symbolizes the greatness of the United States as a "land of opportunity." At the same time, Buckhart is able to slip back and forth across the borderline of either society; however, it is clear that he is at home on the Anglo-American side. His identity as an Apache is recognized by most characters in the series without any explanation offered as to how they know this. Since westerns were the most popular dramatic genre in U.S. television during late 1950s, the series' writers must have presumed that viewers would have accumulated a modicum of historical knowledge, albeit rewritten and romanticized: Apaches had been extraordinarily fierce in their resistance to Euro-American encroachment; subduing them was a monumental challenge and therefore a significant chapter in U.S. history. Another likely explanation is that that strangers—as well as viewers—recognize Buckhart's headband or hairstyle as signifiers of his Apache-ness. A more likely explanation, however, has to do with intertextuality: Ansara had previously spent two years as Cochise on *Broken Arrow*, which ended its run on ABC only two weeks before *Plainsman*'s NBC debut.[11]

Like Tonto, Buckhart appears to have no relatives; he never mentions or reminisces about his family. Not having family is a common convention in savior or "avenging-angel" figures, including the Lone Ranger, Tonto, Batman, and others. Family responsibilities might hinder an avenging angel's work and deter his risky behavior and sacrifices. The closest thing Buckhart has to kin is a white foster daughter whom he rescues after her father is killed by white robbers in episode 1. She lives in the same boardinghouse as Buckhart and is mostly reared by its matronly owner.

Buckhart's hypermasculine physicality is a large component of his persona. Ansara was a very tall man with a large, athletic frame. His costume (he wears the same outfit in nearly every episode) is custom tailored to accentuate his physique. Ansara's body and displays of it are a significant aspect of this series; indeed, when he is in full-Apache mode his body becomes a spectacle. As Richard Dyer points out, in the 1950s bodily displays were considered feminine; it was unseemly for a white man to display his body in this manner.[12] In two episodes, Buckhart spends a considerable amount of time shirtless, and in a third he is literally forced at gunpoint to strip by outlaws who fear miscegenation (they think their sister is romantically interested in him, and in this they are correct; episode 26, "Stella").

This aspect of the character seems to have some relation to William Rice Burroughs's Tarzan, and indeed, Buckhart looks much like Tarzan in his loincloth. In this shot, Buckhart is at an obvious disadvantage, yet he is situated on the left side of the screen, which in this series is an indicator that he will prevail. The light lands on his upper torso. In episode 3 ("A Matter of

Life and Death") Buckhart also doffs his shirt, undergoing a transformation in which he becomes another, far more powerful being. Like Clark Kent/Superman or Dr. David Banner/the Hulk, this ritual involves removing his outer garments, signaling a transformation. Indeed, his superhuman aspects are even noted in the text: in episode 13 ("Toll Road"), an antagonist alludes to Buckhart's superhuman sensory perception—"I thought Apaches heard better than *regular* human beings." Buckhart exchanges his western riding boots for moccasins and crawls around on his hands and knees to sneak into a villain's tent, where a woman is preparing to give birth. Buckhart's "Injun skills" are augmented by his elite education, which positions him to diagnose the woman's septicemia and save her life.

Thus Buckhart combines the best features of both races. He embodies Native American physicality and spirituality combined with Euro-American intellect and education. As in *The Lone Ranger*, intellect and ethics are represented as European qualities and physical prowess and instinct as Indian qualities. Intellectually, he is fully assimilated—a true believer in western values with his illustrious education in Enlightenment science and philosophy along with his training in Anglo-American jurisprudence. At the same time, his Injun skills enable him to solve cases no white man could hope to crack. Buckhart easily straddles this mind/body dichotomy by combining both. In this sense, he is a superman, complete with secret identity.[13]

OPENING SEQUENCE

Using well-established genre conventions from both film and television westerns, *Plainsman*'s opening sequence provides viewers with a great deal of background information in a very efficient manner. The opening sequence has much in common with *The Lone Ranger*'s: we see a man on a horse traversing the "wide-open spaces" of the American West. An important aspect is the reification of the landscape. Both the Ranger and Buckhart are set against verdant hills, lending them strength, with wide vistas of unspoiled land conspicuously absent of human habitation. This raises the concept of *terra nullius*, or land that is unused and therefore free for the taking. As in *The Lone Ranger*, the opening shows a man on a white horse galloping at full speed, with blaring horns in the background. The shaky (handheld?) style here gives the impression of immediacy and truth (the establishing shot of Buckhart is taken from an approaching helicopter, whereas the opening shot of the Ranger is taken from a truck running alongside him). It is necessary at this point to ask, whose gaze does the camera's point of view represent? The point of view from the helicopter suggests that the viewer is looking down from the heavens as God himself might. Thus, before the series even begins, the viewer is positioned (or "constructed") as omniscient and provided with esoteric, hidden information about the character.[14] The implication is that Buckhart is being looked upon—or looked after—by God. Thus the camera work from the very first scene flatters Euro-American viewers by positioning them as God's people. Buckhart is God's messenger, an angel sent to look after the chosen people of the West, the Euro-American settlers.

The character's direction changes from one scene to the next, violating the 180-degree rule. This too works to position the viewer as omnipresent, able to see from two directions at once (figure 4.1). A medium identifying shot of Buckhart shows his face. This shot emphasizes his long hair and Apache headband, letting the viewer know he is a very different kind of lawman. Yet he wears European-style clothes and rides western, not bareback like an Indian; he also carries a sidearm on his right thigh. He appreciates and admires the technological advances of Enlightenment Man. Buttressing the religious symbolism is Buckhart's white horse, which is an ancient symbol of divinity.

Figure 4.1. Helicopter shot from *Law of the Plainsman*'s opening sequence moving in slowly.

VISUAL ANALYSIS

Buckhart is the latest incarnation of the ancient avenging-angel figure who appears in the Old Testament books of Daniel, Malachi, Ezekiel, and Isaiah as well as in the New Testament book of Revelation.[15] This series follows the avenging-angel formula that is so popular in American literature, especially in westerns. For instance, each episode begins with an establishing shot of a peaceful plaza in a small, Edenic village. According to Jewett and Lawrence, nearly all avenging-angel stories start with a community of innocent people whose idyllic existence is disturbed or threatened by an evil force that has to be dealt with by way of righteous violence, which ordinary people do not know how to deliver. Of course, most dramatic formulas begin with an equilibrium that is disturbed by some alien force and must be restored by the protagonist; the variation here is *that only an avenging angel* is capable of administering redemptive violence.[16] Other visual clues reinforce this avenging-angel image.

Buckhart is often shown riding his white horse, his figure outlined against the sky. He appears as a divine figure, with clouds serving as his halo (figure 4.2). Employing the opposite approach of the opening sequence, wherein the

Figure 4.2. Buckhart framed against the sky with clouds for a halo.

viewer is positioned as godlike, this shot is notable for its low angle, setting the character against a background of sky: Buckhart, instead of the viewer, is now godlike. The pine tree in the upper left forms a sort of crucifix, reinforcing the character's status as a biblical-style avenging angel or sacrificial savior.

The manipulative aspects in the character's framing and proxemics are apparent. Much of his screen time is spent in close-up, which has the effect of ingratiating him with the audience, highlighting his sincere, often thoughtful expressions. Until *The Lone Ranger*, canonical westerns tended to depersonalize American Indians. Camera angles, close-ups, and editing techniques affect the way viewers feel about a character. Close-ups were usually reserved for important characters, who were almost always white. This constant scrutiny creates a "positive affect" for the viewer: the more close-ups a viewer sees of a specific character, the more likely he or she is to empathize with him or her.[17] Buckhart's face is examined frequently, signaling that he has nothing to hide—he is almost always sincere and affable. Close-up shots mostly occur when the character is in his "white" or cerebral mode, and full shots displaying his body occur when he is in his "Indian" or physical mode.[18] Buckhart is often shot from slightly low angles, accentuating his height and imposing stature. In terms of framing and composition, Buckhart is frequently positioned in the upper-left sector, indicating his power—except when his control is threatened. Here Buckhart is in a position of authority, in the upper-left sector, shot from chest height. He is "facing off" with a competing law-enforcement officer who turns out to be a bogus sheriff. The key light here, indicating a point of scrutiny, is on the villain's face and his stolen badge. Yet his position, touching the right side of the frame, indicates his imminent demise. He is being pushed out of the frame by the force of Buckhart's righteous indignation. Tables turned: the next shot alerts the viewer that Buckhart's control is threatened and that power relations between the two characters have been reversed. This time Buckhart is being "squeezed out" of the frame. Indeed, Buckhart does get shot—but not killed—later in the episode.

The next shot appears to be taken from the eye level of the townsfolk who are confronting Buckhart, accusing him of not doing enough to protect the town from marauding outlaws. Even though Buckhart towers over those confronting him, his position on the right side of the frame indicates his control is threatened or diminished, and his face is partly obscured. Yet Buckhart still looms large in the foreground, and the key light, emanating from the upper left, is on his face. The shadow cast by the hat over the man's face who is shouting at Buckhart could easily have been removed with a fill light; therefore, one must assume that this was intentional and designed to indicate his being "in the dark" about that which he speaks. He is soon revealed to be an unenlightened racist. In the next shot not only is Buckhart on the right side of the screen, he is well to the lower-right, indicating he is threatened, even though he appears nonplussed. The source of the key light now emanates from the other side of the room, its mobility providing another cue that the camera's (hence the viewer's) "transcendental gaze" is both omnipresent and omniscient: not only does it see everything it needs to see, but it "knows" exactly where to look.[19] The bars of the stair rails envelop Buckhart's head, suggesting he is trapped. The line of the handrail points directly to the ex-army officer who is threatening to lead a vigilante group to usurp Buckhart's authority as lawman. The townsfolk are literally lined up behind the army officer, ready to support his plan and his authority.

NARRATIVE AND HISTORICAL ANALYSES

In narrative terms, this confrontation is very similar to one in Delmer Daves's 1950 film (as well as in the 1956 TV series of the same name) *Broken Arrow*: a member of an "enlightened" branch of the federal government—in this case the U.S. Department of Justice; in Tom Jeffords's case, it was the Bureau of Indian Affairs—has a confrontation or standoff with a more atavistic member of a different service, in both instances an army officer. This is a centrist view that supports the federal government: right-wing "extremists" are bad people; more so than the Apaches themselves. It also serves to assert the moral superiority of the president's *appointees* over the more entrenched, bureaucratic military: thus the federal government is not monolithic in its views—nor its prejudices. This could also be an example of "damage control": it acknowledges that there are indeed racist individuals in the government (holdovers from an earlier era) while younger, more idealistic individuals like Buckhart appear in order to rectify wrongs and fight for change. This would have been an important subtext for the civil rights era as well as for voters in the upcoming presidential election (1960 was an election year).

There is some historical basis for the Buckhart character. Such officers did exist; a notable one was Sam Sixkiller, a Cherokee reservation police officer who doubled as U.S. marshal and prison warden in 1875.[20] Sixkiller, however, operated almost exclusively on the Cherokee reservation, whereas Buckhart operates in a white public sphere, usually against white villains.

In *Law of the Plainsman*, Buckhart is rarely treated disrespectfully by the nicer townsfolk; they mostly see him as an ordinary immigrant—despite the fact that *they* are the immigrants and he is the native. Buckhart is an accepted member of mainstream (white) society, yet he can revert to being an Apache when the need arises. A few unenlightened citizens—almost always criminal types or lowlifes—harass or disrespect him because he is an Apache, but most of the townspeople seem to like him and, more importantly, accept him as one of their own in the Great Melting Pot that is America. Buckhart has done his best to "integrate." He quietly endures racial slurs from unscrupulous whites. In episode 22 ("A Question of Courage"), a bartender refuses to serve him—this could be a veiled comment on the segregation practices that were still prevalent in the Deep South in 1959. In this series, only bad people and fools harbor racism and prejudice, an important message supporting the civil rights movement. It is useful here to refer again to Gray's three discursive strategies used in portraying minority characters, *pluralist* ("separate but equal"), *assimilationist* ("integrated"), and *multiculturalist* ("diverse"). In 1954, with the *Brown v. Board of Education* ruling, the federal government moved from a pluralist to an assimilationist stance in which blacks were to be "integrated" into mainstream (i.e., white) society.

However, most American Indians preferred the pluralist vision and sought to maintain their separate status.[21] *Broken Arrow*, which appeared two years after the *Brown v. Board of Education* decision, reflects the pluralist approach that had prevailed since the *Plessy v. Ferguson* ("separate but equal") decision of 1896: the Chiricahua Apaches enjoy their own ostensibly sovereign spaces, and Cochise remains their leader (even though his white "blood brother," Indian Commissioner Tom Jeffords, is the ultimate but benevolent authority on the Apache reservation). The Apaches are equal to the extent that they are allowed to maintain their own cultural practices and folkways: they are granted *limited* autonomy as long as they stay on the reservation and do not threaten or hinder the dominant group's interests. So even on *Broken Arrow*, "separate" did not equate to "equal." By 1959, however, the discursive strategy, as exhibited in *Plainsman*, had shifted from pluralist to assimilationist. *Plainsman* clearly reflects the new, official paradigm set forth by the *Brown* decision. Gray describes this new assimilationist approach as "appeal[ing] to visions of color blindness, similarity, and universal harmony":

> Such series consistently erase the histories of conquest, slavery, isolation, power inequalities, conflicts, and struggles for justice and equality that are central features of US society. . . . To the extent that these tensions and conflicts are addressed at all, they figure primarily through *individual characters* with prejudiced attitudes, who then become the focus of the symbolic transformation required to restore narrative balance.[22]

Racism is not erased, but as Gray explains, it is depicted as a result of *individual* prejudices and shortcomings rather than as systemic. In *Plainsman*, educated characters unquestioningly accept integration/assimilation as the norm. Racist acts are perpetrated by bullies and right-wing, war-loving ignoramuses. This strategy displaces structural racism—and crimes against humanity—onto unredeemable individuals. This coincides with the federal government's need to rehabilitate the United States' racist image in the international arena.

Buckhart himself rarely makes an issue of his Indianness: he usually lets racial insults pass without reaction. Although he is occasionally depicted as righteously indignant, he is rarely angry. Only once does he appear agitated about how his people have been treated. In episode 9 ("The Gibbet"), a prominent citizen in a nearby town within his jurisdiction is organizing a lynch mob to avenge a neighbor's murder. Once again, Buckhart tries to dissuade the townspeople from taking the law into their own hands. The leader of the lynch mob says, "You don't know what it's like—I've seen a dozen men die senselessly." Buckhart explodes, "I *do* understand—I saw my whole *nation* die senselessly."[23] The fact that the word "nation" is used—a term that implies sovereignty—indicates that the writers have some idea of the legal implications involving treaties with Native Americans. This is an extraordinary statement when compared to most pre–World War II depictions of "Injuns" as irrationally hostile. Buckhart does not expound on what he means, but this could be interpreted as an indictment—of sorts—of U.S. Indian policy and seems to fit with an emerging theme in Hollywood, clarified in Ford's *Cheyenne Autumn* (1964), that U.S. treatment of Native Americans was overly harsh and therefore "senseless." The meaning of this term "senseless" should be carefully weighed. Was the slaughter senseless because it was inhumane—or merely *unnecessary* because the Apaches had already surrendered? If the latter is the case, then, such senseless killing was not unjustifiable, merely uncalled for. Thus the violence may have been *excessive* in its application, justifiable but administered less than judiciously. On the other hand, Buckhart himself could be making an admission that his people fought beyond the point where fighting served any rational purpose. Is he suggesting, then, that the Apaches were the senseless ones? These issues are left unresolved and ambiguous. Probably both meanings are in play.

Episode 23 ("Dangerous Barriers") exhibits another of the rare instances in which Buckhart takes a stance on Native Americans' rights and race relations. In this episode, a Mexican Spanish family is in danger of losing its land to aggressive settlers applying strong-arm tactics in an effort to force the family to sell. Buckhart tries to protect the old Mexican man and his son from these racist thugs. The marshal, Andy Morrison, explains that the Mexican family has been in the area for two hundred years and that its land grant came from the king of Spain. Morrison unwittingly reiterates the legal device of *terra nullius*—that the land was devoid of human habitation and therefore unclaimed, an erroneous assertion Buckhart is quick to correct:

> **Morrison:** When they [the Mexican family] come out here, there wasn't nothing—nobody here (violins play a tense, tremolo chord connoting suspense).
>
> **Buckhart:** (indignant) *My people* were here!

Morrison apologizes, explaining that there has been too much violence between the three groups (Apaches, Spanish, and Anglo-Americans):

> **Morrison:** Wouldn't say *your* people helped much, either.
>
> **Buckhart:** (shouting) My people were at war!

This is interesting in that the series at this point openly takes the position that the Apaches were defending themselves from Euro-American encroachment, rather than the canonical view that they attacked innocent settlers for no apparent reason other than innate hostility or the prospect of plunder. Later, Buckhart has a talk with one of the Anglo settlers who is after the Mexicans' property. The settler, employing typical "producerist" rhetoric, dismisses all previous claims to the land on grounds that it was not being efficiently exploited:

> **Buckhart:** They [the Mexicans] have been here a long time, Mabry.
>
> **Mabry:** Long don't count. They had the chance, and they never done nothin'. You wanna know who made this country? Us. Men like me, comin' out here when we was kids—cuttin' back mesquite and chapparal, scratchin' for water, gamblin' everything we had on a few head of cattle, workin' in the winter till our hands froze and in the summer till our skin burned clear through. Don't talk to me about long.

Mabry reveals he is a white supremacist, saying to Buckhart, "I don't like your kind [Apaches] no better than his [the Mexicans]." By limiting racist animosity to criminals and lowlifes, *Law of the Plainsman* suggests that racism and hatred were not endemic, but rather the products of ill-educated, maladjusted individuals. The historical record shows that this was not the case at all.[24] Buckhart, the mediator, is caught in between the lowlife white supremacists and the Mexican family. The young Mexican man, heir to the estate, also hates Apaches, explaining that an Apache war party raided his family's home and killed his mother and several others. However, after seeing that Buckhart is prepared to sacrifice himself for the cause of justice, the young man redeems himself—the required transformation has been made and equilibrium restored. Episode 23 takes a positive stance by disavowing the concept of *terra nullius* as well as the specious producerist argument (i.e., the previous inhabitants were not making good use of the land and were therefore hoarding it), replacing these with a new, melting-pot vision in which all three groups are to coexist. This corresponds to Gray's *multicultural* strategy.[25] The fact that the Apache and the Mexican align against venal white settlers does not seem to present a problem, since the Apache (Buckhart) is an official representative of the Anglo-American power structure, the federal government, which is here to provide justice for all—even for despised minority groups. No mention is made of African Americans—they are never represented or even mentioned anywhere in this series—but it is reasonable to assume that Buckhart speaks for them as well. This series' ideological project, then, seems largely to be the valorization of the executive branch and its officials, a traditional project of the Hollywood film studios. If the viewer accepts Buckhart as God's earthly representative—as suggested in the opening sequence—then his chief, the president, would be the equivalent of God. Indeed, thirty months prior to this episode's airing, President Eisenhower had sent U.S. marshals (and National Guard troops) to protect African American students attending Little Rock's heretofore segregated Central High School.

Episode 28 ("Jeb's Daughter") deals with the issue of miscegenation. The Anglo-American settlers of Santa Fe are in an uproar because they think Buckhart intends to marry an itinerant, uneducated white woman whom he is actually protecting from her father, who is literally trying to sell her. This woman is so uneducated she is like a savage. The father keeps her tied to a rope, like a beast. Yet the settlers presume that a nonwhite man's marrying a white woman—no matter how low class or uneducated she might be—is unthinkable under any circumstances. The issue of miscegenation is derailed by the issue of class: her social status, despite her color, is lower than Buckhart's, proving that an elite education results in higher status regardless of race. Education is the main marker of class in this series. Buckhart and the redneck settlers are very different, not only because he is an Apache and they are white, but also because he is educated and they are not. Lack of education thus comprises the character defect that fuels racism and intolerance. Buckhart, however, is not judgmental of this wild woman. Even though he has a relatively high social status, thanks to his Harvard education and his secure job, Buckhart retains his democratic ideals: having been deemed a savage himself, he knows how it feels to be in her position, looked down upon by so-called civilized society. He even tries to teach her proper manners and how to dress, much like the character Henry Higgins in the popular musical *My Fair Lady* (which was then appearing on Broadway).

THE ULTIMATE DEMONSTRATION OF THE REGULATOR'S LOYALTY

It is not enough for Buckhart to become fully assimilated into Anglo-American society by defecting and becoming an enforcer of the dominant group's norms. He must continually *prove* his loyalty, and the best way of doing so is by killing one of his own. In episode 6 of *Plainsman* ("Blood Trails"), Buckhart must bring in Charlie Wolf, a "renegade" Apache who has killed a U.S. marshal and a deputy while escaping from jail—and he must do it alone. Because he can think like an Apache—after all, he used to be one—Buckhart is the only person capable of getting close to Charlie Wolf. During this tedious and trying process, Charlie Wolf tries to humiliate Buckhart, calling him a "tame Apache" and a traitor (a fink), and he has a point: Buckhart has indeed turned against his own people. Buck Heart and Charlie Wolf had been boyhood friends. Buckhart must now use his Injun skills to capture and fight mano a mano with Charlie Wolf, who refuses to be taken alive. Buckhart finally kills Charlie Wolf, exactly as Cochise did to renegade Apaches who broke the treaty he made with General Howard in *Broken Arrow* (episode 1, "The Mail Riders"). In killing one of his own, Buckhart has performed the ultimate demonstration of his loyalty to the white man's law.

It must be remembered that *Law of the Plainsman* is a Cold War text. After World War II, European-style colonialism had become discredited, yet the French were still trying to hang on to Algeria and Vietnam, the latter with U.S. assistance and funding. The British had already demonstrated the benefits of "neo-colonialism," which operated not so much through occupation and direct application of force—which could still, however, be retained as a last resort—but by recruiting or even creating a class of collaborators among the colonized.[26] It was necessary, as Biskind explains, to identify a "Third Force," a mediating group, who might be used as a wedge, who would willingly work for the colonizers' interests and even internalize these as their own.[27] This is exactly what Cochise does when he executes renegade Apaches who violate his treaty with the whites. Buckhart goes even further: whereas Cochise lives in an ostensibly separate-but-equal Apache space, Buckhart abandons Apache life to live in town, with all the amenities and privileges that come with being a respectable bureaucrat. What is more, he is truly dedicated to Anglo-American jurisprudence—enough that he is willing to kill one of his own to enforce it. As Biskind notes, the tactic of recruiting collaborators and making them leaders was

> a way of making sure that those who were turned would stay turned—as leaders they now had a stake in the system that made them leaders . . . a reward for coming in out of the cold. . . . The carrot that made the stick unnecessary

> was simply the right to dream the American dream, with some assurance that at least part of the dream would be fulfilled.[28]

Buckhart himself explains that he wanted to be on the side that was winning and enjoy the fruits of collaboration and leadership. According to his account, certain (very astute) Apaches saw that their people were being decimated and realized if they defected they might acquire good educations and good jobs and live in nice houses instead of wickiups.

MODEL MINORITY?

Contrary to the idea, suggested earlier, that American Indians were used as stand-ins or surrogates for African Americans, it is worth considering the alternative possibility that they might also have been used at this juncture as a "model minority" in contradistinction to African Americans. According to Gary Mar, the term "model minority" first appeared in print in 1960 in the *New York Times* and in *U.S. News and World Report*, while *Law of the Plainsman* was still on the air. This trope was originally applied to Japanese and Chinese Americans,[29] but it might also have applied to American Indians. Some scholars go as far as to suggest that this model-minority stereotype had the effect—intentional or otherwise—of discrediting claims of African American civil rights advocates by "proving" that the United States is the land of opportunity where anyone can succeed if only he or she works hard enough.[30] This specious argument has been used to demonstrate that there is no race discrimination in the United States and instead blames its victims. According to scholar Frank Wu, the model-minority trope "conceals within it an invidious statement about African Americans along the lines of the inflammatory taunt: 'They made it; why can't you?'"[31] The implication is that blacks simply have not worked hard enough, and if their position in society is unequal it must be their own fault. After all, if Asian Americans or American Indians can "pull themselves up by the bootstraps," African Americans should be able to do so as well. Perhaps American Indians here were unwittingly being used as a model minority, with African Americans, by comparison, viewed as the "failing minority."[32] Buckhart fits this ideological mold: he has worked relentlessly to improve his station in life. In addition this appears to be the case in the 1974 series *Nakia*, in which Navajo deputy Nakia Parker's boss exhorts that the police should "[h]ire more Indians" (episode 12, "Pete"). The model-minority trope reappeared during the Reagan years (1981–1989). This trope might be deployed whenever blacks appear to become too demanding—or perhaps whenever stratification in U.S. society becomes too obvious.

THE INDIAN AS "WANNA-BE WHITE MAN"

Sam Buckhart is paradigm of virtue, like the Lone Ranger, practically if not literally a saint. The Native American savior—of which Buckhart is the first—was an unusual figure, yet if the series' ratings are any indication, it was hardly accepted by the viewing public. Offering a change from Hollywood's Bloodthirsty Savage stereotype of the 1930s and 1940s, *Law of the Plainsman* was probably seen as quite progressive in its day, perhaps too much so. It promotes tolerance by depicting the United States as the Great Melting Pot, where all who work hard enough—and who thoroughly internalize the dominant group's values—can succeed. *Plainsman*, however, works a bit too hard in making a case that Euro-Americans are not *all* racists—only a scattered few here and there—and that most really did not mean any harm, even if they did spout racist rhetoric from time to time. Even good old Marshal Morrison—a substitute for President Eisenhower—has moments when he ignorantly iterates racist rhetoric, but his *heart* is in the right place; he is simply a bit uneducated. Most such characters come to realize the error of their ways. This probably served as a positive example for white children during the civil rights era. However, it also serves to imply that racism was the product of deviant individuals rather than the entire society.

The (retrospective) impression *Law of the Plainsman* provides, that the political mainstream (i.e., center) was antiracist, is thoroughly misleading. As Richard Drinnon thoroughly demonstrates in his historical study of the links between racism and westward expansion, presumptions of white supremacy went largely unquestioned from the nation's very beginnings and were certainly the norm in the 1880s.[33] Thus these characters do not really represent historical figures at all but rather are metaphorical stand-ins for *contemporaneous* figures such as Eisenhower and his federal marshals. *Law of the Plainsman* flattered 1950s audiences by allowing them to distance themselves from racism, reassuring them that Americans are basically good people despite a few "bad apples." Euro-American viewers apparently needed reassurance that their predecessors did the right thing taking the land away from the original inhabitants—yet apparently they still felt some discomfort about the brutal methods that were employed, especially after World War II, when Euro-Americans became horrified at what the Nazis had been doing in Europe.[34] Euro-Americans did not wish to be reminded by the Soviets that they had committed similar crimes. As a Cold War text, *Law of the Plainsman* served to counter Soviet accusations in the international arena as well as to reassure ordinary Americans that these accusations had little or no merit. Chirot and McCauley explain, "[Euro-]Americans reduced the dissonance between their professed ideals and the reality of the immense suffering they imposed on American Indians by claiming it was 'for their own good.'"[35]

Buckhart (the Regulator), then, is here to reassure the Euro-American public that American Indians do not really mind after all, that in the long run the Anglo-American way was better for all concerned.

NOTES

1. Davidson Loehr, *America, Fascism, and God: Sermons from a Heretical Preacher* (White River Junction, VT: Chelsea Green, 2005), 88.
2. *Law of the Plainsman* also ran as a summer replacement on ABC in 1962.
3. Ansara also appeared in two episodes of *The Lone Ranger*, although not as an Indian.
4. Peter Biskind, *Seeing Is Believing: How Hollywood Taught Us to Stop Worrying and Love the Fifties* (New York: Henry Holt, 1983), 10–13, 18, 32, 41, 48.
5. Ibid., 10.
6. Such pairings often include an "ethnic sidekick"; see Hernan Vera and Andrew Gordon, *Screen Saviors: Hollywood Fictions of Whiteness* (Lanham, MD: Rowman & Littlefield, 2003), 154–84.
7. Tricia Jenkins, "Nationalism and Gender: The 1970s, *The Six Million Dollar Man*, and *The Bionic Woman*." *Journal of Popular Culture* 44(1), February 2011, 101, 108–10.
8. Thomas Borstelmann, *The Cold War and the Color Line: American Race Relations in the Global Arena* (Cambridge, MA: Harvard University Press, 2001), 75.
9. Actually this merging had begun formulating in U.S. literature with James Fenimore Cooper's Natty Bumppo and Chingachgook.
10. Chief Mingo of the Cherokees in *Daniel Boone* (1964–1970) also has an elite education: this half-English son of Lord Dunmore, governor of the Virginia colony, was educated at Oxford.
11. *Broken Arrow*'s last telecast was on September 18, 1959; *Plainsman* made its debut on October 1.
12. Richard Dyer, *White* (London: Routledge, 1997), 146.
13. *The Adventures of Superman*, starring George Reeves, was a popular TV series at this time.
14. Aumont et al., 1992, 214.
15. Book of Revelation 6:2, New American Standard Bible (Grand Rapids, MI: Zondervan, 2011).
16. Robert Jewett and John Shelton Lawrence, *The American Monomyth* (Lanham, MD: University Press of America, 1988), 169–74.
17. Joshua Meyrovitz, "Multiple Media Literacies." *Journal of Communications* 48(1), Winter 1998, 96–108.
18. This dichotomy and how it is resolved would also be crucial in the 1993 series *Walker, Texas Ranger*, in which character Cordell Walker becomes Indian at will, which gives him supernatural powers, much like Clark Kent's becoming Superman.
19. Thus the camera becomes the eye of God.
20. "Sixkiller Assassinated: Cherokee Chief of Police Riddled with Bullets." *New York Times*, December 27, 1886, 1.
21. Vine Deloria Jr., *Custer Died for Your Sins: An Indian Manifesto* (Norman: University of Oklahoma Press, 1988; originally published 1969), 172–73.
22. Herman Gray, "The Politics of Representation in Network Television." In Horace Newcomb, ed. *Television: The Critical View*, 6th ed. (Oxford: Oxford University Press, 2000), 294–95; emphasis added.
23. Buckhart's "whole nation" didn't actually die—the Chiracahua Apaches who survived their clashes with the U.S. Army were by 1877 confined to reservations.
24. Drinnon, 1980, 232–42.
25. This was a fairly new concept in the 1950s, one that may have originated with Elliott Arnold's 1947 novel *Blood Brother*.

26. Britain employed this method using sepoys, or native enforcers, in its conquest of India in the 1850s. There are numerous other examples: Belgian colonizers had done much the same in the Congo by selecting Tutsis for administrative positions; Algerian Jews sided with French invaders against the Muslims, and so forth.

27. Biskind, 1983, 165, 173.

28. Ibid., 165–66.

29. Gary Mar, "Are Asians Model Minorities?" *Race, Racism and the Law*. University of Dayton, 1993. http://academic.udayton.edu/race/01race/model01.htm.

30. Tamar Lewin, "Report Takes Aim at 'Model Minority' Stereotype of Asian-American Students." *New York Times*, June 10, 2008. http://www.nytimes.com/2008/06/10/education/10asians.html.

31. Frank Wu, *Race in America beyond Black and White* (New York: Basic Books, 2002), 49.

32. Stacey J. Lee, *Unraveling the Model-Minority Stereotype* (New York: Teacher's College Press, 1996), 126.

33. Drinnon, 1980, 232–42.

34. Angela Aleiss, *Making the White Man's Indian: Native Americans and Hollywood Movies* (New York: Praeger Publishers, 2005), 62, 69–71, 76.

35. Daniel Chirot and Clark McCauley, *Why Not Kill Them All? The Logic and Prevention of Mass Political Murder* (Princeton, NJ: Princeton University Press, 2010), 54.

Chapter Five

Mingo

Cochise of Kentucky in Daniel Boone *(1964–1970)*

> *[These] images were developed as responses to problems, as symbolic means of reducing the complexities and ambiguities of the American situation to a simple, satisfying formula. . . . [They] replace the troublesome and problematic facts of the real world with a counterworld of pseudofacts. The reader is thus given a new world to perceive, more consistent than the real one and more in line with his own hopes, fears, and prejudices.*—Richard Slotkin[1]

Four years after *Law of the Plainsman* failed to hold the interest of the viewing public, NBC, in collaboration with Twentieth Century-Fox and Arcola Productions, unveiled *Daniel Boone*. These Revolutionary War–era stories are very loosely based on the real-life exploits of a former colonel in the Kentucky militia (which was then part of the British colony of Virginia), who after the revolution became a hunter-trapper and land surveyor.

This chapter will introduce and examine the trope of the Indianized white man, which has been popular in U.S. literature (and later film and TV) since Cooper's Natty Bumppo, who was loosely based on the real-life Boone. This Indianized white man remained a popular figure all the way to 1950 (and 1956) with Tom Jeffords of *Broken Arrow* (this does not include hybrids and "half-breeds" such as Hondo Lane, John Rambo, Billy Jack, and Cordell Walker). The chapter will also further explore issues of the land and who controls it, which is, after all, what is at stake in most westerns, even if this is glossed over. *Daniel Boone*, however, deals with issues of the land in a surprisingly forthright manner. This chapter will also examine how American Indians are represented in the series, that is, which Indians are "good," which are "bad." Finally, it will consider the series' nationalist message in light of the Cold War and in defining the United States itself: was it the leader of the Free World, as U.S. leaders claimed, or was the United States just another flagging empire, as the Soviets claimed?

THE INDIANIZED WHITE MAN AND ANGLO-AMERICAN MASCULINITY

The Indianized white man is a key trope in understanding westerns, as well as most stories about relations between whites and American Indians. It is also a key element in the foundations of Anglo-American (i.e., white, heterosexual) masculinity in U.S. mass media. In the post–Revolutionary War period (1790s–1800), a new myth was required for the new nation. According to Richard Slotkin, the leaders of the newly liberated colonies were looking for a prototype male who would be uniquely "American," as opposed to European.[2] He would have to possess American Indian qualities yet still be distinct from—and superior to—Native Americans themselves. The solution was to synthesize two apparent opposites: this "New Man" would have European (i.e., Enlightenment) intellectual capacities, but he would also have American (i.e., native) physical qualities and instincts. He would be a rugged man, comfortable in the wilderness, a loner, an individualist, unencumbered by civilization and its pretenses.[3] Thus the Indianized white man took form. Since even before the Revolution, Euro-Americans found this image both attractive and useful.[4]

The theme of the Indianized white man has long been a recurring—and useful—one in literature addressing white encounters with Indians. Todorov cites, for example, the story of conquistador-turned-native Gonzalo Guerrero, who, after being shipwrecked off the coast of Mexico in 1511, defected to the Mayans and, in the manner later employed by T. E. Lawrence (and the fictional Lieutenant John Dunbar of *Dances with Wolves*), taught them how to defend themselves against white encroachment.[5] Guerrero became a full member of the group, marrying a "high-ranking" maiden with whom he had several children. When invited by Cortez to return to Spanish civilization, Guerrero refused and was eventually killed by Cortez's forces in 1528. An early colonial British version of the Indianized white man was the story of Thomas Morton, who in 1627 was accused by Plymouth Colony governor William Bradford of sponsoring bacchanalian-style festivities with Indians. In 1637, Morton published a mocking critique of Puritan America titled *English New Canaan*. Another example was Major Robert Rogers, leader of Rogers' Rangers, who in 1766 published a play titled *Ponteach [Pontiac]: Or the Savages of America* based on his experiences as a British soldier, which also was critical of Puritan society and sympathized with, even romanticized, Indians. Rogers rather than Boone might have been adopted as the national myth-hero had it not been for his Loyalist (pro-British) sentiments.

Richard Slotkin writes, "It was the figure of Daniel Boone, the solitary, Indian-like hunter of the deep woods, that became the most significant, most emotionally compelling myth-hero of the early republic."[6] In 1827, James Fenimore Cooper created his Indianized hero Nathaniel Bumppo, who was largely based on the legend of Boone created in John Filson's 1784 account.[7] A decade after Bumppo's debut, Robert Montgomery Bird created Nathan Slaughter, a Jekyll-and-Hyde-type character, who transforms himself from a mild-mannered Quaker into the brutal killer known as the Jibbenainosay. The major difference between Bumppo and Slaughter is that the latter hates Indians (Bumppo was himself ambivalent). Slaughter's parents were murdered by the Shawnee; he learns Indian ways only so he can track them down and kill them. "Nathan becomes more savage than the Indians."[8] Both the Cooper and the Bird series are set in Kentucky.

THE INDIAN COMPANION, GUIDE, AND ALLY

Where does the American Indian fit into this myth-making? For one thing, he provides the template American masculinity is largely based upon. As imagined by whites, he has superhuman senses, intuition, tracking skills, and speed. Despite lack of formal education, he is clever and quick. He is also in touch with nature and knows the land intimately. These are qualities the Indianized white man will borrow. In addition, Good Indians like the Cherokee in *Daniel Boone* are quick to grasp and appreciate the superiority of the white man's ways—there is some historical basis for this—and thus provide a sort of *endorsement* through imitation. What is more, the imaginary Indian like Crusoe's Friday, Bumppo's Chingachgook, the Lone Ranger's Tonto, Tom Jeffords's Cochise, and Boone's Mingo serve not only as companions to the white invaders but as guides and enablers—they assist in the ongoing project of Euro-American conquest of the continent.

THE HISTORICAL DANIEL BOONE, MODEL FOR THE NEW MAN

After the American Revolution, the real-life Daniel Boone (1734–1820) claimed a parcel of land for himself and eked out a living as a hunter-trapper. Boone also found he could make money as a guide and surveyor for land speculators. He even did some unsuccessful speculating of his own.[9] Once again, the land and who controls it is the crucial issue.

The real Boone was highly Indianized. In the process of leading settlers to Kentucky, he was captured by the Shawnee in 1778 but became so well respected by them they adopted him as a full member. This has been a recurring theme in literature addressing white encounters with Indians. How many times has this trope—the white man who goes native and teaches the natives how to protect themselves—resurfaced in such films as *Broken Arrow*, *Little Big Man*, *Dances with Wolves*, and *Avatar*? Boone was adopted as a son by Shawnee chief Black Fish, who named Boone "Big Turtle" because he was short and stout—unlike Fess Parker, who was six feet, six inches and rather trim.

Boone's story was recounted in 1784 by land speculator John Filson, who was trying to generate publicity for his project of luring homesteaders to Kentucky in order to sell them parcels of land that had been seized from natives. Filson's 1784 account of Boone's captivity and other adventures struck a nerve: Boone became an overnight legend, the mythical New Man, uniquely American. This figure would remain vastly popular: Cooper's Leatherstocking tales, which created the template for western novels, were inspired by Boone's adventures.[10] Bumppo was, like Boone, a "mountain man," an Indianized "man of the wilderness." This figure became quite prominent toward the mid-1850s. Not only were there several mythicized accounts of Boone's exploits, authors such as Hawthorne, Bird, William Gilmore Simms, and Timothy Flint created Indianized white heroes that used Boone as a touchstone. This literary type—soon to become an archetype—struck a chord in the popular imagination and still resonates in such figures as Cordell Walker (see chapter 8).

BACKGROUND OF THE SERIES

Daniel Boone was developed by Twentieth Century-Fox Pictures in association with Arcola Pictures (Aaron Rosenberg and Marty Melcher, Doris Day's husband), with actor Fess Parker through his production company, Fespar Enterprises. Parker had previously starred as Davy Crockett in three episodes of Walt Disney's ABC-TV anthology series along with two feature-film versions.[11] Crockett, too, was an Indianized white man who had captured the country's imagination via television.

The series' initial episode, "Ken-Tuck-E," aired on September 24, 1964. The TV Boone lives very simply and speaks like an uneducated "hillbilly" (being from Texas himself, Parker had a convincing southern accent). A connection is made here between hillbillies—generally of Ulster-Scots descent—and Indians, especially Cherokees: both rely on their wits rather than "book learning" (however, Boone's friend and companion, Cherokee chief Mingo, has an Oxford education); both are enemies of the English. This, however, is historically inaccurate: the Cherokees *resented* the Scots-Irish and their incursions and sided with the British, who had promised to curb advances by settlers.[12] The role of the Scots-Irish in conquering the continent was recognized by none other than Theodore Roosevelt—Scots-Irish himself on his mother's side—who wrote a four-volume history of the winning of the

West. Roosevelt remarked that if it weren't for the ferocity of these Ulster-Scots—who formed the core of the so-called pioneer movement—the United States would not exist in its current form:

> That these Irish Presbyterians were a bold and hardy race is proved by their at once pushing past the settled regions of America and plunging into the wilderness as the leaders of the white advance. . . . They were the first and the last set of immigrants to do this; all others have merely followed in [their] wake.[13]

Although these settlers are never explicitly identified in the series as Scots-Irish, there are clues: for example, many of them, including Boone, speak with a distinct hillbilly "twang." One of the militia leaders, Captain Ives (Robert Lansing) actually speaks in a distinct Scottish brogue (season 1, episode 1, "Ken-Tuck-E"). Although Boone is depicted as a hillbilly, the historical Boone was neither Scots-Irish nor Presbyterian: he was born in Pennsylvania of English and Welsh stock and raised a Quaker. The TV Boone, in any case, has more in common with Indians than with the British, who refer to the settlers as "ragtag rebels" (season 1, episode 12, "Not in Our Stars," air date December 31, 1964). Unlike the British, who are seen as invaders, Boone somehow "belongs" to this land. He is therefore a natural enemy of the British foreigners. Not only that, he is a commoner who, like the Indians, believes in equality for all men; the British are part of the old feudal hierarchy.

VISUAL AND NARRATIVE ANALYSES

The opening sequence clearly tells the viewer that Daniel Boone is a model of American masculinity. The first words of the series' theme song go,

> Daniel Boone was a man
> Yes, a big man
> With an eye like an eagle
> And as tall as a mountain was he[14]

Thus the viewer is told in no uncertain terms that American masculinity is what is at stake in the program. However, these lyrics are historically inaccurate: Boone himself was not especially tall.[15] These words are accompanied by a close-up of Boone, then a medium shot of him aiming a rifle. Masculinity here is connected to firearms.

There is a cut to Boone's point of view showing a black bear running directly toward the camera; Boon, standing tall, confidently aims and fires (figure 5.1). Next the camera shows Boone hiding behind a tree aiming and firing his rifle. This is immediately followed by a shot of three armed American Indians running toward the camera and dropping dead (figure 5.2).

Figure 5.1. Boone readies to shoot a charging bear.

This is remarkable for two reasons. First, it is misleading: Boone himself was not, as he was often portrayed, an avid "Indian fighter"; in fact, as a Quaker he preferred to avoid bloodshed whenever possible.[16] Second, the scene with the Indians charging echoes the previous scene of the wild bear charging; Indians are thus equated with wild animals. This posits a blatantly dehumanizing implication. Shooting a bear is part of taking control of the territory: the bear is an obstacle to progress—synonymous with Manifest Destiny, willed by God.[17] By the same token, the Indians comprise a natural obstacle and, like bears, must be removed in order to clear a path for western civilization and commerce. However, since World War II, it was unusual to show Euro-Americans killing Indians in films and even more unusual on network television. After all, these series were seen overseas, and the state department found it inconvenient for Hollywood to continue fomenting images of Euro-Americans slaughtering and forcibly relocating indigenous peoples. As discussed in the introduction, this was something the state department would prefer to see revised or at least elided. The implications of this opening montage did not go unnoticed; indeed, it would be revised and moderated in

Figure 5.2. Indians charging (from Boone's point of view).

later episodes. The new version began with a loving long shot of the land with a river running through it.

Boone is positioned in all the predictable ways reserved for a man of the wilderness and as a god. He is visually connected with a mountain, symbolizing strength. Subsequently he is framed against the sky in a godlike pose (figure 5.3). He is looking up at the sky as if receiving a message from the heavens. He is God's messenger. The theme song continues,

> Daniel Boone was a man
> Yes, a big man
> And he fought for America to make all Americans free

This, of course, is a literal reference to Boone's role as a Revolutionary War militia leader. What he is fighting for is America's freedom from British rule. However, the historical Boone did not fight to make *all* Americans free: only white Americans of European descent who had the wherewithal to acquire land. The actual freedom that was being fought for in this region at the time—and as plainly seen in this series—was the freedom of white speculators to wrest land from Native Americans and sell it to immigrants, in this

Figure 5.3. Boone framed against the sky in a godlike pose.

case, mostly Scots-Irish settlers. This freedom to treat the land as a commodity was being impinged upon by the British.

That the land is the primary issue here is made obvious in the visuals and is dealt with openly in the text, which seems unusually forthright for a western. Many episodes open with long shots lovingly depicting the land as if it were a beautiful woman (figure 5.4). The land is the prize in most westerns: ownership and control of it is what these texts are mostly about. The series echoes the *producerist* view of the land: the land truly belongs to the people making the best use of it. Boone's early sidekick, Yadkin, makes the following remark in the context of the British coming to remove the Boonesborough settlers: "If fightin' off Injuns, clearin' the land, raisin' cabins, sowin' the crops ain't valid title for a chunk of land, I just don't know what is" (season 1, episode 12, "Not in Our Stars"). In another episode, Boone echoes this doctrine, telling a British officer: "The only people who own a country are the people who live on it—people who plow the ground, plant the orchards, and learn to love it" (season 2, episode 16, "Gabriel"). This would obviously preclude nonsedentary or nomadic Indian groups. Mingo asserts that the Cherokees were the original landowners of Kentucky, and apparently they made the deal with the settlers. Lord Dunmore, Mingo's father, howev-

Figure 5.4. Shot of the landscape from a later version of the series' opening sequence, replacing the Indian-shooting scene.

er, disagrees: "The Cherokees never did own Kentucky"; according to Dunmore, Kentucky had been Tuscarora land. In any case, the settlers urge violence. Boone, however, counsels patience and urges all to work within the system: "The law's gonna do our fightin' for us, not powder and shot." Thus Boone, unlike the Lone Ranger (discussed in chapter 2), endorses the legal system, as does Mingo.

ENTER THE REGULATOR

Boone's friend and ally, Cherokee chief Mingo is played by Ed Ames (real name Edmund Urick), who appeared in the series from its debut until he left in 1968. Mingo is a hybrid. He is the son of a Cherokee mother and an English father, John Murray, earl of Dunmore, who happened to be the governor of the Virginia colony. Mingo was schooled at Oxford. Sam Buckhart (of *Plainsman*) also had an elite education, having been schooled at Harvard. Like Buckhart, Mingo possesses extraordinary physical and intellectual capabilities. Mingo's English is perfect—far better than Boone's, in fact. Mingo is more educated—and therefore more "white"—than Boone, while Boone is as Indianized as any Indian. There is a sort of "trading places" going on here. Mingo could easily pass for English—which he does when he impersonates a British officer (season 3, episode 11, "Requiem for Craw Green"), yet—unlike half-breeds Hondo Lane or Cordell Walker—he favors and cherishes his Indian side.

For all his education and erudition, however, Mingo still represents a regression toward the Tonto/Friday mold. Like Tonto, he plays "second fiddle" to the white hero. Even though he is the nominal chief of the Kentucky Cherokees, he takes directions from the Anglo-American protagonist, who is generally—although not always—the better tactician and better decision maker. Like Tonto, Mingo is the "junior partner" in the enterprise of American expansion. Mingo, for his part, knows the land and its geography intimately as well as its array of indigenous peoples, some of whom, like the Tuscarora, fight for the dastardly British. He serves as Boone's Indian scout and informer. In fact, in one episode, a Creek woman calls Mingo a traitor (season 3, episode 7, "The Matchmaker"). Mingo, however, is not quite as obsequious as Tonto, who was more or less the Ranger's manservant; Mingo is closer to Cochise in *Broken Arrow*: an ostensibly autonomous leader who finds it prudent to defer to the white hero's better judgment. As Kim Trevathan notes, Mingo "pretty much went along with whatever TV's Boone proposed."[18] This is curious because it would seem that the native would know the geography better than a white man. But Mingo has been raised mostly in the white man's world, and besides, a white man is better at all things Indian than Indians are.[19] In "The Matchmaker" Shawnee "princess" Little Fawn wants to choose her own husband, disregarding the custom of arranged marriage supposedly practiced by her people. "The white man's way is better," she says. This is indicative of the *endorsement* feature of programs featuring American Indians: these programs very often depict fantasy Indians endorsing Euro-American norms; in fact, this may be one of their primary functions. The white man is treated as infallible, which no doubt serves to flatter mainstream, male viewers. When Boone asks Little Fawn's father, Chief Two Shoots, what he is supposed to do with the girl, the father replies, "You are the white man—you will know." The white man, after all, can handle any situation.

After the great assimilationist leap embodied in Sam Buckhart in *Plainsman*, Mingo takes the Indian character a step backward toward what Gray calls the pluralist, or separate-but-equal doctrine embodied in the 1896 *Plessy v. Ferguson* decision. This is perhaps an indication that the integrationist/pluralist strategy may not have been attractive to the public—and this might have been demonstrated by the failure of *Law of the Plainsman*, a failure production executives most likely did not wish to repeat—even though *Plainsman* had embodied the official federal government stance since the *Brown v. Board of Education* decision. The networks, then, not only had to placate the government as well as advertisers—they had to please viewers, too, which proved even more difficult.

Daniel Boone recognizes that Indians were divided into various competing groups, not all of them opposing white settlement. The Cherokees seem to be on the side of the (presumably) Scots-Irish settlers, while the Tuscarora, whose leader, Tamend, is portrayed as a red-coated buffoon, openly side with the British ("Not in Our Stars"). However, in real life, the Cherokees supported the British. Only the Shawnee act independently. Boone himself is an honorary Shawnee and wears the leather necklace (*quippo*) designating him as such.

Mingo is a contradiction. He has been educated in England, and his English is better than most whites' including Boone's. He even sings opera in Italian. He also satirizes the Indian stereotype, speaking in faux-pidgin English using an accurate Tonto imitation (season 4, episode 5, "Beaumarchais"). Mingo eventually makes the ultimate display of loyalty to his white brothers: he kills his half-Creek twin, a Bad Indian who has shot a settler, in much the same manner that Sam Buckhart in *Plainsman* kills his childhood friend Gray Wolf (*Law of the Plainsman*, episode 6, "Blood Trails"). In another episode, Mingo takes up the battle for a young Cherokee who has fought with the Continental Army and was rewarded with a land grant by the new federal government (season 3, episode 11, "Requiem for Craw Green"). Yet greedy, land-grubbing settlers are trying to drive him off this land, too:

> **Mingo:** Hmmm—Indians being driven off their own lands.
>
> **Kanati:** My father is a brave man, Mingo. He will not be driven off the land that belongs to him.
>
> **Mingo:** But does it belong to him, Kanati, legally?

Mingo cuts his hair, passes as white, and becomes an avenging angel, taking it upon himself to right this inequity. As in *Broken Arrow*, this series acknowledges that unscrupulous speculators and settlers were engaged in taking land from Native Americans; however, these unscrupulous operators are differentiated from those who, like Boon and his clients, acquire the land "legally." Thus the entire system is not questioned, only a few "bad apples." Moreover, the question of legalities, first presented in *The Lone Ranger*, again arises: legal according to whose laws? This episode reflects what really happened to the Cherokees of Georgia, who adopted many of the white man's customs: Cherokees registered land claims, built houses, adopted a written language and a system of laws and law-enforcement, even converted to Christianity. Yet none of this was enough for them to be recognized as human beings with basic human rights when white settlers wanted their land.

This series clearly sympathizes with the Cherokees. Tuscaroras, on the other hand, are ridiculed and reviled as tools of the British, while the Creeks and Shawnees are portrayed as mysterious, unpredictable, and dangerous. Why Cherokees are chosen here to be portrayed as allies of the rebels—since the reverse was actually the case—is an interesting question. Certainly, if the new nation needs to be symbolically endorsed by American Indians, *some* native group had to fulfill that function.

THE WEST AND THE CONCEPT OF UPWARD MOBILITY

Since all the land in Europe—and most if not all of the Eastern Seaboard—had been claimed, and men rigidly hierarchized into castes, the only option for young men who wanted to carve their way in the world was to go west. It was the same in any colonialist country: young men went to the colonies to wrest fortunes from lands seized from indigenous peoples who lacked the technologies to defend them. There was land for farming, cutting timber, mining, or reselling to settlers—even for raising and selling farm animals and slaves. For some it was the lure of gold and quick riches. For others it was simply the promise of a small patch of land to cultivate and feed a family. Andrew Jackson was one of the industrious—and ruthless—ones who wrested his fortune from the land and its bounties. The West was imbued with the promise of upward mobility. The concept of upward mobility is reflected in the series: Boone himself is, like Jackson, a self-made man of the wilderness. *Daniel Boone* clearly evinces the conservative (perhaps even reactionary) philosophy of "bootstrapping": any man willing to endure hardship, work hard, and pull himself up by the bootstraps can succeed. "Success" here means upward mobility. Seen by viewers in the 1960s, an age of burgeoning consumerism based on class, this would have been a welcome message, validating and reinforcing their own desires for upward mobility through product purchases.[20] It would also illustrate the moral superiority of American-style laissez-faire capitalism over the burgeoning welfare state during a period when the "evils" of socialism were being hotly contested.[21]

CONCLUSION

Daniel Boone served to propagate one of the foundational myths of U.S. society, that the new country was a revolutionary state founded on freedom for all. However, the historical record clearly demonstrates that this freedom was only for landowning whites and mainly referred to freedom of commerce. The primary freedom depicted in the series was the freedom of land speculators to sell parcels to desperate settlers who wanted to engage in subsistence farming and feed their families. In the series this land-speculating business is depicted as questionable, even perhaps intrinsically unfair to the Indians—but as long as it is done according the letter of the law, it is deemed acceptable. In any case, it is—like Manifest Destiny itself—unstoppable.

While the United States is depicted as the land of freedom for all, Britain, on the other hand, is referred to as an empire, even by British officers themselves, who use the term proudly. These Kentucky settlers—proto-Americans—are averse to this word, which in this context assumes a connotation of totalitarianism. This would be true of most viewers in 1964 as well. The words "imperialism" and "empire" had become dirty words in the United States: empires are ipso facto evil. This attitude would become even more pronounced in the 1980s when President Reagan reignited the Cold War, dubbing the Soviet Union "the Evil Empire"; if the USSR is an empire, and if the United States is its structural opposite, then the United States cannot be an empire.[22] In fact, it was born as a revolutionary state (this fact is debatable and will be discussed later). The term "empire" can only apply to those decadent European powers, the British, the French, the Spanish—certainly not to this new nation about to be conceived.

By 1964, the term "freedom" had become a metonym for all things American; in fact, the United States was considered not only the leader but the heart of "the Free World." U.S. television during the Cold War served and promoted these nationalist ideals.[23] The Cold War was still going strong in 1964, when Boone debuted. Lyndon Johnson, a fervent Cold Warrior, was president. The military would soon implement Johnson's "escalation" of the Vietnam conflict into a full-blown war, which eventually put five hundred thousand U.S. soldiers in Southeast Asia. Thus it might be worth considering that Boone, like the Lone Ranger, personified the ideal U.S. leader or perhaps the United States itself. Frontier themes, which had been extremely popular during the Eisenhower presidency, had begun to resurface in the form of President Kennedy's "New Frontier." Westerns, however, were losing popularity at this time; Kennedy's New Frontier was more likely to be in outer space than in the Old West. Despite this—as well as the tendency for many new series (mostly police dramas) to be located in the "urban frontier"—*Daniel Boone* remained a phenomenally successful series, running a total of seven seasons.

Daniel Boone purported to be historically based—yet the Kentucky legislature passed a resolution condemning its inaccuracies in 1964.[24] The original Boone was a loner, a solitary figure whose isolation in the wilderness makes him stronger. But in the series Boone never travels alone. It is, of course, a storytelling convention to provide the protagonist someone to talk to, to confide in; hence the addition of sidekicks and companions (Tonto wasn't part of the original *Lone Ranger* series; he was added later to give the Ranger someone to explain things to). The series invented a sidekick (Yadkin, who only lasted a year and was largely replaced by Mingo) in order to facilitate the mechanics of storytelling. But why was Boone's companion an Indian, and why specifically a Cherokee? Perhaps there are psychic issues related to American Indians and their subjugation that were being "worked through." If Mingo actively cooperates with the Euro-Americans, then the Euro-Americans seem less racist, willing to incorporate Indians into the creation of a new empire—and in fact the Cherokees did cooperate with whites, not that it ultimately did them much good.

What is more likely is that the Boone character, like the Lone Ranger, personified America itself and that the Indian character, like Tonto, is a junior partner in the global project of Manifest Destiny (see discussion in chapter 2 on *The Lone Ranger*). The suggestion here is that our "little brown brothers"—whoever these might be—should join with Euro-Americans to fight evil, totalitarian empires. It is ironic that the empire depicted here would become the United States' real-life "junior partner" in globalization: Great Britain.

Thus the series was not just entertainment but a rhetorical device—in other words, propaganda—designed to sell Cold War ideology. This would not be surprising in light of the fact that Twentieth Century-Fox (TCF) was committed to producing propaganda (TCF's former vice president of production Darryl F. Zanuck's own term). What is more, NBC was at this time closely associated with the federal government, what Senator Fulbright might have called an "adjunct of government" (or what Althusser would call an Ideological State Apparatus). Its parent company, RCA, was a prominent defense contractor, receiving billions of dollars in revenues from the government. As a 1961 internal memo made clear, NBC executives were well aware of the benefits of selling the Cold War agenda and its concomitant anxieties to a nervous public. NBC executive Harry Bannister was coming up with ideas about how the Cold War could be used to sell air conditioners, washing machines, radios, and such. What is more, Bannister noted that amplifying the threat of atomic war "could make us an awful lot of friends in Congress and in the government."[25] NBC, owned by RCA, a major defense contractor, was especially attuned to the government's ideological needs. In another example, when Ed Ames (Mingo) left the series in 1968, Boone's new sidekick would be an escaped slave, Gabe Cooper, played by former football star Roosevelt Grier. This followed closely on the heels of the 1968 Kerner Commission report, created in the wake of an epidemic of urban uprisings, which recommended that television roles for African Americans be expanded.

NOTES

1. Richard Slotkin, 1973, 467.

2. Ibid., 294–306.

3. Hamid Naficy and Teshome H. Gabriel, *Otherness and the Media: The Ethnography of the Imagined and the Imaged* (Langhorne, PA: Harwood Academic Publishers, 1993), 46. Also see R. W. B. Lewis, *The American Adam: Innocence, Tragedy, and Tradition in the Nineteenth Century* (Chicago: University of Chicago Press, 1955), 41, 175, 201.

4. Philip J. Deloria, *Playing Indian* (New Haven, CT: Yale University Press, 1998), 2–9.

5. Tzvetan Todorov, *The Conquest of America: The Question of the Other* (New York: Harper and Row, 1984), 195–96.

6. Slotkin, 1973, 21.

7. Ibid., 512.

8. Ibid., 514.

9. John Mack Faragher, *The Life and Legend of an American Pioneer* (New York: Henry Holt, 1993), 174, 247, 261.

10. Slotkin, 1973, 23–24; K. Randel Jones, *In the Footsteps of Daniel Boone* (Winston-Salem: John F. Blair, 2005), 13; Frederick Zackel, "Robinson Crusoe and the Ethnic Sidekick." *Bright Lights Film Journal* 58, November 2007. http://www.brightlightsfilm.com/30/crusoe1.html.

11. Parker, who had neglected to negotiate a participation in ancillary Davy Crockett items featuring his likeness, such as bubble-gum cards, lunchboxes, etc., was careful to negotiate a 30 percent ownership stake in *Daniel Boone*, in which he wore a coonskin cap like the one he had worn in his role as Crockett ("Him Mingo," 1967). Disney had already created a Daniel Boone episode (air date December 11, 1960) for his long-running anthology series.

12. Antipathy between the Scots-Irish and the British went back as far as the Test Acts of 1704, in which the Ulster Presbyterians (Scots-Irish) had been prohibited by the British from practicing their religion.

13. Theodore Roosevelt, *The Winning of the West*. Volume 1, *Spread of the English-Speaking Peoples* (New York: G. P. Putnam's Sons, 1910), 127.

14. Vera Matson and Lionel Newman, "Theme from *Daniel Boone*" (20th Century Music Corp., 1964).

15. Faragher, 1993, 30.

16. Ibid., 86, 174, 300.

17. On Indians as obstacles, see Drinnon, 1987, 190. Drinnon quotes a 1951 article from the *Journal of Negro Education* that states, "Indians were frequently a real menace to the advancing tide of settlement."

18. Kim Trevathan, *Coldhearted River: An Odyssey down the Cumberland* (Knoxville: University of Tennessee Press, 2006), 33.

19. Churchill, quoted in *Images of Indians: How Hollywood Stereotyped the Native American*, directed by Chris O'Brien and Jason Witmer (Starz Encore Entertainment, 2003).

20. Paul Fussell, *Class: A Guide through the American Status System* (New York: Touchstone, 1983), 25, 84–85, 117–18.

21. There was even a course taught in high schools called "Americanism vs. Communism" with a textbook that included an introduction by FBI chief J. Edgar Hoover.

22. Stanley Lens, *The Forging of the American Empire: From the Revolution to Vietnam, a History of U.S. Imperialism* (London: Pluto Press, 2003), vii.

23. Tricia Jenkins, "Nationalism and Gender: The 1970s, *The Six Million Dollar Man*, and *The Bionic Woman*." *Journal of Popular Culture* 44(1), February 2011, 95.

24. Vincent Cosgrove, "A Tubful of Coonskin and Corn." *New York Times*, October 29, 2006. http://www.nytimes.com/2006/10/29/arts/television/29cosg.html?_r=0.

25. Harry Bannister, memo to NBC president Robert Sarnoff, October 17, 1961. Harry R. Bannister collection, State Historical Society of Wisconsin. Quoted in Watson, 1994, 170.

Chapter Six

Regulator on the Urban Frontier

Hawk *(1966)*

This chapter examines the ongoing transformation of the Regulator from "apprentice white man" to fully integrated member of the dominant society, illustrating Gray's *assimilationist* discursive strategy, which became paramount in the 1960s (discussed in chapters 2 and 3).[1] However, integration tended to erase or eliminate differences between ethnic groups. In the 1970s and beyond, this process would come to be considered akin to genocide in that it erased subcultural differences. As a corrective, cultural differences—or "diversity"—would later be celebrated, not eliminated (what Gray calls a *multicultural* strategy). As Clark's model predicts, during the transition from Non-recognition to Regulation that occurred on television in the 1960s with regard to minority characters, such characters seem to be stereotyped within administrative, military, and law-enforcement positions, not the most prestigious nor lucrative occupations in U.S. society. But this is how minorities prove their loyalty to the society at large.

Following this pattern, New York City police detective Lieutenant John Hawk (Burt Reynolds), who describes himself as a "full-blooded" Iroquois (episode 9, "The Living End of Sisterbaby"), has submerged himself into the Great Melting Pot and in doing so largely jettisoned his identity as an Iroquois. Clearly this follows Gray's *assimilationist* strategy. Unlike 1970s Navajo lawman Nakia Parker, whose Indianness haunts him—and to which he clings—Hawk's Indianness is submerged. Hawk is a step forward in terms of assimilation, while Parker is a step toward diversity or *multiculturalism*, a sort of middle ground.

Another notable aspect of this series is how the "law and order" trope is used to critique the implied lawlessness of uprisings in urban African

American ghettoes. What is more, *Hawk* subtly mocks and derides the burgeoning youth culture as well as Manhattan's "effete" artistic community. The program—like most cop shows of the era—seems to appeal to reactionary viewers and thus is inherently conservative.[2] As we shall see, this conservative constituency who enjoyed crime dramas was one presidential hopeful Richard Nixon would succeed in cultivating (he called them the "silent majority"). However, the traditional cop show mentality is moderated here as Lieutenant Hawk is constructed more as a mediator or social worker than as a police officer. The networks were well aware of their need to cultivate younger audiences, and this they did by infusing programs with "social relevance."[3] *Hawk* tries to have it both ways, with the Regulator as an outsider, while trying not to be too obviously condescending toward young viewers lest the program lose this valuable demographic segment. *Hawk* is vaguely reactionary, together caricaturing bohemian types as dilettantes, drug users, at best unwitting dupes of venal villains. The series walked a tightrope, simultaneously romanticizing and disparaging youth culture; as the textual analyses will demonstrate, this required a great deal of ambiguity. Clearly this program operated in what Barthes called the realm of myth, seeking a "third way" out of an ideological impasse.[4] Yet this was a dilemma no amount of fence straddling or ambiguity was able to finesse. John Hawk, being a romanticized "outsider," functioned as a mediator of sorts between the two generations but nonetheless remained an "Establishment" figure, albeit a somewhat hip one. Just as it takes a Man Who Knows Indians to fight Indians, it takes an outsider to understand—and deal with—deviants.[5]

We have also seen how the native enforcer has taken on some aspects of comic-book hero Superman (specifically his "alter ego" or secret identity). Cartoonist and comics expert Jules Feiffer states that Superman's story is a "tale of ethnics": "Superman was the ultimate assimilationist fantasy."[6] This is a fairly good description of *Hawk* as well. John Hawk also takes on qualities of Britain's Cold War hero James Bond.

INSTITUTIONAL CONTEXT

The year 1959 saw a great leap in the development of the American Indian character on television: the apprentice white man was promoted from ethnic sidekick to full-fledged lawman on *Law of the Plainsman* (discussed in chapter 4). That same year, an American Indian U.S. Treasury agent (William Youngfellow) appeared on ABC's *The Untouchables* (1959–1963).[7] The template was thus established for the Native enforcer of the white man's law: the Regulator. After the failure of *Plainsman*, however, American Indian characters remained obscure for several years (in another period of Nonrecognition), with the exception of Daniel Boone's half-Cherokee, Oxford-educated sidekick Mingo on *Daniel Boone* (NBC, 1964–1968). In 1966, the Indian lawman took a surprising leap into the modern era with *Hawk*. This in itself is exceptional: as the overview in chapter 1 indicates, characters set in the contemporary era comprise only 18 percent of American Indian television characters.

As the western genre faded in popularity, the "urban frontier" began to replace the western frontier. By the time *Hawk* made its debut on ABC, urban crime shows had been popular for some time, with *Dragnet* (NBC, 1951) and *Naked City* (ABC, 1958–1963) being prime examples. If the urban crime program can be seen as a variation on the western, then John Hawk would be the modern equivalent of Sam Buckhart in *Law of the Plainsman*. Like many police programs of its day, *Hawk* is largely anti-urban: Its overall message seems to be that the city is a treacherous place with too many dangerous people, one implication being that small towns and suburbs are more prudent places to live. Yet, at the same time, the program celebrates John Hawk's urbane, sophisticated lifestyle.

BURT REYNOLDS'S BACKGROUND

Actor Burt Reynolds brought many intertextual implications to the role of John Hawk. Reynolds, who claims his father was half Cherokee, had already played several American Indian and "half-breed" characters in television and film; in fact his "ethnic" look helped launch his career.[8] Reynolds had portrayed half-Indian blacksmith Quint Asper on CBS's *Gunsmoke* from 1962 to 1965 and had played Indian warrior Red Hand in a 1965 episode of a short-lived NBC series titled *Branded.* Reynolds himself stated he had already been typecast as a half-breed.[9] "The only Indian I haven't played is Pocahontas."[10]

Just weeks before filming *Hawk*, Reynolds appeared in Sergio Corbucci's spaghetti western *Navajo Joe* (1967), an Italian production filmed in Spain.[11] Reynolds plays a Navajo who is significantly Europeanized and speaks perfect English. His hair is longish but tidy in an expensive-looking Continental "shag" cut. He is clean shaven. He wears Indian buckskins. His appearance suggests a link between American Indians and hippies, and indeed, 1967 would become famous for its burgeoning hippie movement and San Francisco's "Summer of Love." The Navajo people Joe lives among are a peaceful group; his woman is attacked by a gang of outlaws collecting scalps. The gang is white except for its leader who is a vicious half-breed. Joe is Europeanized in that he takes on spiritual qualities normally reserved for the White Savior: he becomes an avenging angel, an instrument of divine retribution. Like Sam Buckhart of *Law of the Plainsman* (see chapter 4), he is the Indian equivalent of the Lone Ranger but far more violent. His task is to administer redemptive violence, and he uses his extraordinary "Injun skills" to assist in this task. Many of the same techniques used to deify the Lone Ranger are seen here as well: Joe is often shot at low angles, framed against the sky with clouds for a halo, possesses near-superhuman powers and instincts, shoots with uncanny accuracy, crawls around in the dirt, and still somehow manages to remain neat and tidy.

VISUAL, TEXTUAL, AND HISTORICAL ANALYSES

Hawk was created by veteran radio and television writer Allan E. Sloane for Screen Gems, the TV division of Columbia Pictures. *Hawk*'s executive producer, Hubbell Robinson, had been vice president of programming at CBS television and was instrumental in CBS's postwar effort to expand television roles for minorities.[12] Its production team consisted of former staff members of Emmy-winning series such as *The Defenders*, *Eastside/Westside*, and *Naked City*. *Hawk* debuted on September 8, 1966. Its fast-moving opening sequence borrows from the documentary "direct-cinema" or cinema verité tradition (without the long takes, however), photographing New York City streets through the windshield of a passing police vehicle, using a combination of grainy (16mm), fast film with available lighting.[13] The opening music consists of generic crime- or spy-series "action" music—very much like the theme from *Mission Impossible*—with stabbing, staccato horns and jagged piccolos suggesting frenetic movement.

John Hawk's race, although barely noticeable in the program, is crucial in making him an outsider. This outsider status helps Detective Hawk solve crimes, since he can obtain information a "straight" cop such as Detective Joe Friday (*Dragnet*, 1951–1970) might not be able to elicit. Police dramas had faded since *The Untouchables* (1959–1963) and *Naked City* (1958–1963), but coinciding with conservative "talking points" focusing on law and order (in reaction to inner-city riots) and the upcoming election campaign of Richard Nixon and Spiro T. Agnew, "Cops were back, and this time they were cool. . . . Such officers were hip, not straight or square . . . but the latent messages were conservative."[14] Because of Hawk's American Indian background, "he occasionally dealt with racism inside and outside the department."[15] However, there is little if any visual evidence of Hawk's Indianness: most comes in the form of dialogue; it is almost as if these references to his race were last-minute revisions to capitalize on Reynolds's intertextual status as an American Indian. For example, in episode 13 ("'H' is a Dirty Letter") Hawk iterates, "I have an ethnic hang-up: I come from a long line of losers." It is difficult to ascertain what he means by this remark: "loser" could connote that his father was an alcoholic (we find out later in this episode that his father was disabled in a construction accident), or he could mean his people lost their struggle for control of their land to Euro-Americans. In any case, Hawk is situated as far from the reservation as possible, both geographically and metaphorically: he has no family members, observes no American Indian traditions, and exhibits few if any Iroquois artifacts in his stylish apartment.

Urbanity was a key concept in *Hawk*, and the program's promotional materials made much of the fact that the series was shot entirely in Manhattan.[16] Detective Hawk works at night, and the look of the program is often noir-ish and foreboding. Exteriors were shot on New York City streets; interiors at Filmways' East Harlem facilities.[17] Hawk's urbanity is emphasized, and his swinging-bachelor lifestyle and stylish wardrobe seem to be partly inspired by Hugh Hefner's "Playboy Philosophy."[18] Ian Fleming's spy character James Bond had also fit this playboy mold. Hawk seems to follow suit: both Bond and Hawk are swinging bachelors with macho manners, good looks, good incomes, job security, respectability, nice apartments, and the most fashionable attire.[19] Both Bond and Hawk attract women without even trying. Hawk has a good job and is thus upwardly mobile. Apparently the key to his success, symbolized by his fashionable clothes, is his willingness to assimilate, to learn the intricacies of the dominant group's laws, and, like Sam Buckhart in *Law of the Plainsman* (discussed in chapter 4) and Nakia Parker (in chapter 7), become a Regulator.

Episode 7 ("The Man Who Owned Everyone") begins with a shot of the sky, suggesting that the viewer-subject is visiting from the heavens or that the story is being told from a godlike perspective. As the camera pans across a busy New York street, Detective Hawk stands in a doorway looking to the sky as if receiving a message. This visual serves to provide the character with a divine mission. Later in the program we hear a criminal being arrested refer to Hawk sarcastically as "Chief." This is apparently a common nickname for American Indians, used by whites with a degree of irony.

Episode 9 ("The Living End of Sisterbaby") begins with a party scene in an uptown Manhattan loft. The first scene is loaded with what was then considered cutting-edge camera work. Much of this—the use of smaller, handheld cameras, fast film, and available lighting—originated of necessity in television news and public-affairs documentaries.[20] Handheld technique, open-frame composition, fast film capable of utilizing low light, and grainy-looking stock were used to give fiction films a journalistic aura of actuality and immediacy. Thus documentary-style camera work came to be considered *authentic* rather than simply cheap.[21] These techniques were still relatively new in 1966 when *Hawk* made its debut. *Hawk* employs these devices as a sort of seasoning, saving them for key plot points and action sequences. After the initial excitement of the party scene, the program settles down to the usual array of two-shots, shot/reaction shots, and over-the-shoulder shots. Unorthodox camera work is thereafter reserved for crucial information or action. The party is crowded with young, slender models in miniskirts and fashionably dressed young men dancing "the frug" and "the jerk." It is a notably diverse group: a well-dressed black couple, several young hipsters, an Asian woman, a WASP-ish Congressman and his wife, a young, extremely slender, pale-skinned British model (probably a nod to 1960s fashion icon Twiggy), and two white, working-class uniformed officers standing guard. Hawk comprises yet another flavor in this Great Melting Pot. The writers lump together several disparate groups: hipsters, art-scene dilettantes, fashion models, British mods, and blacks together into a monolithic youth culture or what later came to be called "counterculture." Detective Hawk speaks their language; his Indianness enables him to go undercover among the urban natives, so to speak. The young model says he is "pretty groovy." Yet his attitude and remarks about this group exhibited later in the episode are condescending.[22]

Detective Hawk's personal style appears to be a combination of James Bond, Hugh Hefner, and Bruce Wayne (Batman).[23] Emphasizing his "uptown" urbanity, Hawk spends nearly the entire episode wearing a tuxedo or a custom-tailored trench coat (figure 6.1). The bow tie and the trench coat help him maintain his image as an authority figure. His hair is slicked back, in stark contrast to most of the male partygoers who sport floppy Beatle cuts (Reynolds's toupee is rather obvious also).

Looking quite straight, he nonetheless—a bit self-consciously—adopts African American "hipster" slang in order to ingratiate himself with the youth scene. As a mediator, Hawk seems to have a foot in both camps; what is more, he recognizes this: the "Indian scout" to whom he sardonically refers is a double agent of sorts. The viewer can never be sure if Hawk truly "digs" these bohemian types or disdains them; this ambiguity is no doubt purposeful, for the benefit of both younger and older viewers, to allow each its own interpretation.[24]

Then there is the issue of Hawk's race: on one hand, it seems the bohemians of the worlds of art, fashion, and theater are more likely to relate to Hawk because he is an outsider, as are they. His outsider status encourages counterculture types to trust him and helps solve cases. Yet, like any enforcer or sepoy, he must be very careful not to "go native" and forget which side he is on; after all, there is a job to be done.

Hawk's Indianness, however, is barely noticeable: his swanky uptown apartment contains no visible Iroquois artifacts other than a pair of snowshoes tacked to a wall and a tom-tom fashioned from animal hide, which he keeps on his coffee table. The tom-tom is brought out to link him with the young model he is trying in a very paternalistic manner to rescue from the hipster scene that has been taken over by greedy, murderous hustlers. Hiding out in Hawk's apartment, the model, who is pale-skinned with long, blond hair ("white on white"), merrily taps on Hawk's tom-tom, not realizing that it

Figure 6.1. The debonair Detective John Hawk is much like a James Bond–type character. Being utterly assimilated, he evinces almost no Indianness whatsoever.

signifies his Iroquois heritage (figure 6.2). Since they have not discussed his race, she most likely assumes this is simply a bongo drum, an anachronistic item once used to accompany beatnik poetry.

She most likely also assumes he is a fairly hip, older guy.[25] There is an implied irony (and criticism of her generation) here: this ostensibly hip young person is displaying her ignorance. There is no condemnation of these young hipsters; rather they are depicted as extremely naïve and as a result are being "conned" and manipulated by villains. This is starkly reminiscent of a remark FBI director J. Edgar Hoover made in 1961 that beatniks were unwitting dupes of communists.[26]

Other than the tom-tom and the snowshoes, there are few references to Hawk's Iroquois heritage, and these mostly occur in dialogue, spoken by Hawk himself: for example, he jokingly calls himself an "Indian scout," which is an ironic reference to a Good Indian (i.e., a "friendly") who helps the army hunt down Bad Indians (usually from a different group)—in other words, a double agent, a traitor, a fink. Hawk overtly capitalizes on his Indianness when he is flirting with—or being flirted with by—white women,

Figure 6.2. The young girl's whiteness is emphasized as she taps on Hawk's tom-tom.

almost as if to say, "I'm an exotic delicacy; come taste me." This invokes the common stereotype that nonwhites are more physically than intellectually oriented and therefore endowed with more sexual prowess than the average white male. This is depicted in a nightclub scene in which a very pale chanteuse asks Hawk to "come back and see me—without the badge." The singer's whiteness is accentuated by her platinum hair (like Marilyn Monroe's), and the lighting is carefully arranged to highlight her fair skin (figure 6.2). Compared to her, Hawk is dark, swarthy. Hawk later employs what I call the "flying Indian leap," when he takes a running jump and scuffles with the villain who tries to shoot the young model with his CIA-style camera gun.[27]

There is a hierarchy of races in this police department. If Hawk is second class, answering to a white superior, then Hawk's own ethnic sidekick, African American Detective Dan Carter (Wayne Grice), must be third class.[28]

HISTORICAL CONTEXT

To more thoroughly understand this program, it is necessary to look deeply at the issues and politics of its period. In 1966, an epidemic of race riots was breaking out in urban areas all across the United States (it also appeared that some young whites supported this rebellion and seemed prepared to join in). Conservative politicians began invoking the term "law and order" in their rhetoric, and many television programs echoed this. The phrase, which would be reiterated ad nauseam by candidate Richard Nixon during his 1968 election campaign, was a veiled metaphor for suppressing black—as well as hippie-youth—rebellion.[29] Nixon himself had privately confided that the issue of "law and order" was directly tied to "the damned Negro-Puerto Rican groups."[30] What was the role of American Indians in this rhetoric? They may have been used—unwittingly—as a "model minority." As in *Nakia* (chapter 7), American Indians, along with Asians, may have been portrayed in a "positive comparison" to blacks; indeed the Indian character in *Hawk* is here to fight *crime*. No black criminals are seen in this program, but the word "crime" would become associated in public discourse with a black "underclass."[31]

Hawk lasted only one season on ABC (ten years later, after Reynolds's subsequent breakthrough as a film star, the program was rerun on NBC as a summer replacement). It may be that the U.S. viewing public was still not receptive to the idea of Indian-as-lawman. On the other hand, there may be simpler explanations for the program's quick demise: *Hawk* was poorly written, not well realized, and not especially entertaining: *New York Times* TV critic Jack Gould called the program "excessively melodramatic" and wrote that its "camera technique got the better of [its] characterizations."[32] Another explanation was offered by an anonymous reviewer on Internet Movie Database who felt the main problem was that the character John Hawk "was an arrogant jerk. . . . It was impossible to like him."[33] Reynolds himself, however, felt that the program was well conceived and executed.[34]

CONCLUSION

Despite its contemporary setting and John Hawk's sophisticated urbanity, *Hawk* did not constitute a breakthrough for American Indian characters. Like the Apache character Chief on *Garrison's Gorillas* (1967–1968), the urbane Hawk had been so thoroughly assimilated into mainstream society that his Indianness was scarcely noticeable unless Hawk himself pointed it out in dialogue. Programs like these, in which American Indian characters, largely abandoning their own identities, internalize and enforce the dominant group's norms, conform to what Fanon described as "the final liquidation, indeed the digestion, of the native."[35] Conforming to Clark's framework, minorities must not only assimilate but must make themselves useful to a society that has historically oppressed them.[36]

Despite *Hawk*'s failure, network executives, especially at ABC, did not give up on the image of the Indian lawman. Eight years after *Hawk*'s brief run, Screen Gems, the company that distributed *Hawk*, unveiled a made-for-TV movie, *Nakia*, starring Robert Forster as a Navajo police officer. *Nakia* soon became a weekly program (analyzed in the next chapter) but like *Hawk* lasted only one season. Even in 1974 U.S. audiences were apparently still not receptive to the concept of the Indian lawman. As this study has demonstrated, there had been several such figures, all of which ultimately failed to capture the public's imagination. The concept of the Indian lawman would not take hold with U.S. audiences until this figure was presented as *half* Indian: *Walker, Texas Ranger* (CBS, 1993–2001; discussed in chapter 8) would be the first and only long-running program to feature an Indian lawman in a starring role. In *Hawk*, assimilation had gone too far, erasing differences between races. The pendulum would swing back, reaching a sort of homeostasis in what came to be called *multiculturalism*. Unlike Detective Hawk, who jokes about his Iroquois heritage, Navajo policeman Nakia Parker is deeply concerned with maintaining his Navajo identity.

NOTES

1. Herman Gray, "The Politics of Representation in Network Television." In Horace Newcomb, ed. *Television: The Critical View*, 6th ed. (Oxford: Oxford University Press, 2000, 294–95.

2. Michael Parenti, *Make-Believe Media: The Politics of Entertainment* (New York: St. Martin's Press, 1992), 2–3; Philip D. Smith and Kristin Natalier, *Understanding Criminal Justice: A Sociological Perspective* (Thousand Oaks, CA: Sage Publications, 2004), 161.

3. Erik Barnouw, *Tube of Plenty: The Evolution of American Television* (Oxford: Oxford University Press, 1990), 430.

4. Roland Barthes, *Media in Advertising* (New York: McGraw-Hill, 1964), 129.

5. Richard Slotkin, *The Fatal Environment: The Myth of the Frontier in the Age of Industrialization, 1800–1890* (Norman: University of Oklahoma Press, 1998), 96.

6. Jules Feiffer, "The Minsk Theory of Krypton," *New York Times Magazine*, Dec. 29, 1996, 14–15.

7. The Youngfellow character, a full-blooded Cherokee, played by Abel Fernandez, was based on a real-life treasury agent William Jennings Gardner, who was half Chippewa, raised on a reservation and trained at the Carlisle School in Pennsylvania. *The Untouchables* was based on a memoir by former special U.S. Treasury agent Eliot Ness. Youngfellow, a minor character who stayed in the background, could be considered the conceptual progenitor of police detective John Hawk of *Hawk* (1966) as well as of the character Chief in *Garrison's Gorillas* (1967). Actor Fernandez, who was of Yaqui-Mexican descent, played many American Indian as well as Latino roles in films and on television.

8. Burt Reynolds, *My Life* (New York: Hyperion, 1994), 96–97; Joan Barthel, "Honest Injun." *New York Times*, July 24, 1966, 77.

9. Reynolds, 1994, 129. The *half*-Indian archetype is extremely significant and deserving of deeper study. In 1971, a half-Navajo vigilante was popularized by Tom Laughlin in the movie *Billy Jack*, whose story line is similar to that of *Navajo Joe*'s. Rambo in *First Blood*

(1982) would follow the *Billy Jack* formula as a half-Indian ex–Green Beret gone amok. Pat Aufderheide, "Good Soldiers." In Mark Crispin Miller, ed. *Seeing through Movies: A Pantheon Guide to Popular Culture* (New York: Pantheon Books, 1990), 106. Also see Sam Pack, "Constructing 'the Navajo': Visual and Literary Representation from Inside and Out." *Wicazo Sa Review* 15(1), Spring 2000, 137–56. Like Billy Jack and Rambo, Cordell Walker of *Walker, Texas Ranger* was both half Indian *and* a martial-arts expert.

10. Reynolds, 1994, 138.

11. There was a 1915 Hollywood film short titled *Navajo Joe*, director unknown.

12. "Urge CBS, ABC, NBC-TV to Treat Negroes Fairly." *Jet* 22(10), June 28, 1962, 62.

13. Reynolds, 1994, 111, 118.

14. Smith and Natalier, 2004, 161.

15. Marty MacKee, "Plot Summary for *Hawk* (1966)." Internet Movie Database. http://www.imdb.com/title/tt0059992/plotsummary.

16. Tim Brooks and Earl Marsh, *The Complete Directory to Prime Time Network and Cable Television Shows* (New York: Random House, 2003), 514.

17. Reynolds, 1994, 119.

18. In 1962, *Playboy* publisher Hugh Hefner began writing his "Playboy Philosophy" in irregular installments of the magazine, outlining his concatenation of socially liberal politics, sexual liberation, aesthetic hedonism, and mass consumerism (Hefner 1962).

19. In this respect, Hawk, as a sort of "fashion plate," foreshadows the "MTV cops" of *Miami Vice*.

20. These techniques began in France's cinema verité and were used in the United States, most notably by Robert Drew and his protégés, Richard Leacock, D. A. Pennebaker, and Albert Maysles, in the ABC public affairs program *Bell and Howell Close-Up* (1960–1961). Many of these devices, combined with fast panoramas ("zip pans") and jump-cut edits, were borrowed (or simulated) by big-budget filmmakers such as Richard Lester in *A Hard Day's Night*, 1964. Coming full circle, these techniques made their way back to television—notably in *Star Trek* (NBC, 1966–1969), *The Monkees* (NBC, 1966–1968), and *Rowan and Martin's Laugh-In* (NBC, 1968–1973)—and would later become an essential factor in "crime shows" as well as later "reality shows."

21. Perhaps the best example of the use of handheld technique combined with an inexpensive look giving fiction a sense of actuality was *The Blair Witch Project* (1999).

22. This concept of police infiltration by "hip, undercover cops" would be further explored in ABC's *Mod Squad* (1968–1973), which consisted of two young white hippies and a black quasi-revolutionary.

23. The ABC series *Batman* appeared in 1966, the same year as *Hawk*.

24. In the days before there were dozens of U.S. networks catering to all kinds of tastes, appealing to a niche audience was unthinkable. Each network wanted the largest possible general audience and was therefore compelled to appeal to all demographic segments simultaneously. This required a great deal of ambiguity.

25. Hawk is chided about his age and called "Dad" by a younger cop, who looks and behaves very much like Kookie (Edd Byrnes), a young detective in *77 Sunset Strip* (1958–1964).

26. Roger Chapman, *Culture Wars: An Encyclopedia of Issues, Viewpoints, and Voices*. Volume 1 (Armonk, NY: M. E. Sharpe, 2010), 295.

27. The gun mounted in the camera is a realistic spy reference: in the 1960s, the Technical Services Division of the Central Intelligence Agency developed a .22-caliber pistol mounted in a camera, dubbed "The Stinger." Allan A. Swenson and Michael Benson, *The Complete Idiot's Guide to the CIA* (New York: Alpha Books, 2002), 101.

28. Cordell Walker (*Walker, Texas Ranger*) also has a black sidekick, Ranger Jimmy Trivette, played by Clarence Gilyard Jr., who had been Andy Griffith's ethnic sidekick on *Matlock* from 1989 to 1993. Apparently there is a hierarchy of sidekicks with Indians second to white men, then Asians, then African Americans. Rarely if ever is the situation reversed: I have not seen nor heard of an ethnic hero with a white sidekick except in Mel Brooks' 1974 farce, *Blazing Saddles*. See Hernan Vera and Andrew Gordon, *Screen Saviors: Hollywood Fictions of Whiteness* (Lanham, MD: Rowman & Littlefield, 2003).

29. John E. Roemer, Woojin Lee, and Karine van der Straeten, *Racism, Xenophobia, and Distribution: Multi-Issue Politics in Advanced Democracies* (Cambridge, MA: Harvard University Press, 2007), 60.

30. Joe McGinnis, *The Selling of the President: The Classic Account of the Packaging of a Candidate* (New York: Penguin Books, 1988), 23.

31. Thomas Sugrue, "A Dream Still Deferred." *New York Times* (editorial). March 26, 2011. http://www.nytimes.com/2011/03/27/opinion/27Sugrue.html; Stephan Thernstrom and Abigail Thernstrom, *America in Black and White: One Nation, Indivisible* (New York, Touchstone, 1997), 309.

32. Jack Gould, "TV: NBC Tarzan, He Urbane and Sophisticated." *New York Times.* Sept., 9, 1966, 91.

33. "Burt Reynolds Should Have Played John Hawk More Like Quint Asper and Less Like Ben Casey." *Reviews and Ratings for "Hawk" (1966).* Internet Movie Database. June 11, 2006. http://www.imdb.com/title/tt0059992/reviews.

34. Reynolds, 1994, 118. According to Reynolds's account, *Hawk* garnered "critical raves." Ibid., 122.

35. Quoted in Ward Churchill, *Fantasies of the Master Race: Literature, Cinema, and the Colonization of American Indians* (San Francisco: City Lights Books, 1998), 162.

36. Cedric C. Clark (Syed M. Khatib), "Television and Social Control: Some Observations on the Portrayals of Ethnic Minorities." *Television Quarterly*, Spring 1969, 21.

Chapter Seven

Billy Jack Returns—as a Cop

Nakia *(1974)*

What appears stable and secure—the policeman on the corner, for instance—is only slightly more secure than the white men in the jungle and requires the same continuous but precarious triumph over an all-pervading darkness.—Edward Said (commenting on Joseph Conrad's *Heart of Darkness*)[1]

Media products are complexly regulated mixtures of stimulants and tranquilizers.—Kirby Farrell[2]

This chapter examines the progression of the Regulator's internalization of the dominant group's norms. The Indian Regulator went from apprentice white man in *The Lone Ranger* to full-fledged lawman with white horse and all the trappings of divinity that normally accompany the white savior in *Law of the Plainsman.* By the time John Hawk arrived, the Indian Regulator had completely internalized Anglo-American norms. In examples of Gray's assimilationist discursive strategy, cultural differences are erased in favor of a narrative that conveys America as the Great Melting Pot.[3] Apparently going from one extreme to the other was not an acceptable strategy as far as minorities themselves were concerned. According to Vine Deloria Jr., most Native Americans did not *want* to be assimilated.[4]

Nakia represents the pendulum's swing to the middle, with Navajo police officer Nakia Parker being drawn (back) toward his Navajo roots, questioning Anglo-American values. How can he reconcile these two disparate cultures and their values? First he experiences an identity crisis, after which he is firmly situated on the Anglo-American side but without giving up his democratic ideals, which are conflated with Indianness. Parker's ideas of social justice also rest on elements of working-class status, which are also conflated with Indianness. To him, Anglo culture is too rational. It lacks heart. Issues of class here are blurred with—and obscured by—issues of race, which is reflective of U.S. culture as a whole. Since *Nakia* is a police drama, issues of law and order inevitably come to the fore. This chapter will also examine what the term "law and order" and its related tropes connoted in the program's historical and political context.

Eight years after *Hawk*'s quick demise in 1966, executives at Columbia Pictures-Screen Gems Television, the same studio that produced *Hawk*, decided to give the Indian-lawman character another try. This time he would be a patrolman for the fictional Davis County, New Mexico, sheriff's office. The series would be titled after the lead character, Nakia Parker, a Navajo. *Nakia* (both the made-for-TV movie and the weekly series that followed) was created by writers Christopher Trumbo and Michael Butler. Trumbo's father, Dalton Trumbo, was one of the Hollywood Ten blacklisted by the film industry after the House Un-American Activities Committee (HUAC) Hollywood investigations in 1947. Both Trumbos seem to have been interested in promoting more tolerant views of American Indians: in 1978 Christopher Trumbo and his father worked together on a made-for-TV movie addressing American Indian issues, *Ishi, Last of His Tribe* (the title has an obvious Cooperesque ring).

Nakia's executive producer, David Gerber, had been an advertising executive and a production executive at Twentieth Century-Fox Television (which in 1956 had produced *Broken Arrow*; see chapter 3). Gerber specialized in police dramas with a slight character twist: his police officers were often women or minorities and as such embodied challenges to the traditional view of law enforcement: these were more like *mediators* between city officials and the "little people." This probably had a something to do with meeting the demands of the networks for "social relevance" in order to attract younger audiences.[5] Yet at the same time the networks could not afford to alienate older audiences nor the federal government. Gerber produced *Police Story* (NBC, 1973–1977), *Police Woman* (NBC, 1974–1978), *Joe Forrester* (NBC, 1975–1976), *Today's FBI* (ABC, 1981–1982), *In the Heat of the Night* (NBC, 1988–1995), and several made-for-television movies. Most of these were made by Columbia Pictures Television, of which he became vice president in 1979.

In 1971, Gerber had developed the short-lived *Cade's County*, which contained a minor American Indian character, police dispatcher Joanie Little Bird (as the overview in chapter 1 demonstrates, female American Indian characters are exceedingly rare). *Nakia*, which appeared three years after *Cade's County*, was more or less a continuation of that series with similar settings and sets; the programs even shared one actor, Taylor Lacher, who played a sheriff's deputy in both. *Nakia* was basically *Cade's County* with a Navajo patrol officer in place of a Mexican American: this fits into an earlier observation (in chapter 3) that, in casting terms, Indians and Mexicans are treated interchangeably.

Nakia began as a two-hour made-for-TV movie that aired on ABC, April 17, 1974.[6] It was filmed on location in and around Albuquerque, New Mexico. It would be the second Indian-lawman series set in New Mexico, the first being *Law of the Plainsman*, set in Santa Fe (see chapter 4). *Nakia* featured Robert Forster, of Irish and Italian extraction.

Navajo characters became prominent during the late 1960s and are still one of the most frequently represented Native groups in U.S. media; this may be because they comprise the largest surviving group.[7] Starting about 1966, Navajos became fashionable in films, television, and novels, beginning with Sergio Corbucci's *Navajo Joe*, starring Burt Reynolds (discussed in chapter 6). Elvis Presley, who claimed to be part Cherokee, also played a Navajo named Joe (Lightcloud), in *Stay Away, Joe* (1968).[8] Many Navajo characters have been depicted as Regulators: for example, in the 1970 novel *The Blessing Way*, Tony Hillerman created a reservation-police detective, Lieutenant Joe Leaphorn;[9] Hillerman later gave Leaphorn a partner, Jim Chee, and published eighteen Indian-detective novels, some of which were made into movies. Leaphorn and Chee are reservation-police detectives, while Nakia Parker serves as a uniformed officer in the county sheriff's department (i.e., off reservation), working alongside white patrolmen. In other words, Parker is "integrated" into mainstream society rather than segregated in a Navajo ghetto as are Leaphorn and Chee. This is significant in that integration (i.e., *assimilationist* discursive strategy) was still a theme in U.S. public discourse in 1974. Parker, then, could represent not only Navajos or American Indians but U.S. minorities in general. Although there had already been black characters who had been integrated into police, spy, and space programs (such as Bill Cosby in *I Spy*, 1965–1968; Nichelle Nichols in *Star Trek*, 1966–1969; Greg Morris on *Mission Impossible*, 1966–1973), the issue of integration again became a thorny one due to a 1971 federal-court ruling that mandated busing white children into black school districts. Since advertisers were averse to controversy, the issue of race might have been displaced onto American Indians just as it most likely had been in *Law of the Plainsman* (1959–1960; discussed in chapter 4). Officer Nakia Parker often confronted—and arrested—uneducated, backward white criminals (often colloquially referred to as "rednecks"), some of whom, as seen in *Law of the Plainsman*, resented a minority person having any authority at all. Even in the 1970s, segregationists still felt that minority policemen, especially black ones, lacked the moral authority to arrest a white man.

The pilot of *Nakia* was very different from the weekly series that followed. The series mimicked *Cade's County*, while the made-for-TV movie followed Tom McLaughlin's 1971 film *Billy Jack*, in which a Navajo hero "tries to protect his people from the machinations of villainous whites."[10] In fact, *Nakia*'s plotline was so similar to *Billy Jack*'s that Warner Bros., *Billy Jack*'s distributor, sued for copyright infringement.[11] Billy Jack had been a U.S. Army Special Forces (Green Beret) commando and a martial-arts expert (*Billy Jack*, 1971). These ex–Green Beret, martial-arts, and half-breed motifs later resurfaced in the Rambo film *First Blood* (1982) and in television in *Walker, Texas Ranger* (1993–2001): both John Rambo and Cordell Walker are half Indian. There is very little "backstory" provided here about Parker in *Nakia*; he does not appear to be a trained martial-arts expert. However, he is quick to get involved in fisticuffs and is easily able to beat up three white men at once. This corresponds to the suggestion in many programs that nonwhites are more likely than whites to become physical instead of using their brains. Nakia also engages in the "flying Indian leap," in which he pounces through the air onto his opponents. This was seen in *Law of the Plainsman* (1959–1960) as well as in *Hawk* (1966).

OPENING SEQUENCE

The made-for-television movie's opening sequence is similar to *The Lone Ranger*'s as well as *Law of the Plainsman*'s in several respects: a lone horseman is shown against a beautiful, hilly landscape (the implications of the way the landscape is reified is discussed in detail in chapter 2 on *The Lone Ranger*). Establishing shots of the "wide-open spaces" are tropes used to quickly identify the western genre, although Nakia technically is not a western—westerns are almost always set in a mythical American past—but a modern crime series set in a rural region. It is a hybrid: a police drama with elements of a western.

Although technically a crime drama, *Nakia* demonstrates its debt to the western by visually connecting the hero with the mountains.[12] He is associated with rock formations, nature, and permanence (the Lone Ranger was also associated with a rock formation; see chapter 2). Whereas in both *The Lone Ranger* and *Plainsman* the rider in the opening sequence is going full gallop, Nakia is cantering at a leisurely pace (figure 7.1), suggesting that he is at peace with his environment. It is difficult at first to differentiate the rider from the landscape; he is "connected" to the land, as we later discover. His trek is accompanied by symphonic music in a majestic western style reminiscent of composer Aaron Copland. A prayer, recited in what we later discover to be the protagonist's (Forster's) voice, accompanies the rider's trek up the hillside. This non-diegetic voice intones,

> Great Mystery, you existed from the first
> The sky, this earth, you created
> Great Mystery, look upon me, pity me
> That my people may live[13]

Figure 7.1. Like the Lone Ranger, Parker is associated with rock formations and "touches" the sky.

When the rider recites the words "pity me," he is shot from an extremely long distance, appearing tiny, almost insignificant (figure 7.1). Suddenly, his stature changes: the character goes from insignificant to godlike, exactly as Tom Jeffords does in the opening scene of *Broken Arrow* (1950 and 1956).

He is now seen in a close-up shot, in a godlike pose (figure 7.2). This shot not only lets the viewer get to know the character in close-up, it associates him with the sky, where the Great Spirit resides. The camera then returns to the land, and he appears tiny again. As in most westerns, the land is "empty" (see discussion of this issue in chapter 2). This reification of the land is central to the theme of this made-for-TV movie, even though this is not a traditional western. Still, land is the key issue, and protecting it creates the conflict with which the protagonist must struggle, a quandary for a Navajo whose religion and spiritual beliefs are inextricably bound to the land and its creatures but who at the same time serves as an officer of the dominant culture's law. His attempt to reconcile these opposing views is the crux of this movie, which was used as a pilot for the television series.

Parker is often shown on horseback. In western movies, the "paint" horse is a trope of Indianness. Whereas both the Lone Ranger's and Sam Buck-

Figure 7.2. Parker framed against the sky, his earnest face is given scrutiny in close-up. At the same time he is connected to a rock formation.

hart's horses were pure white—signifying their riders' godliness—Parker's horse, like Tom Jeffords's in *Broken Arrow*, is partially white—connoting the hybridized status of its rider. Forster, the actor who plays Parker, has hazel eyes (seen earlier in close-ups), which reinforces suggestions of mixed ancestry for the character (the movie's story line contains more tension if one views the character as mixed race or half-breed). The character's signifiers are often incoherent and contradictory, some creating dissonance and tension with others. For example, he is Navajo yet at the same time white savior as was Sam Buckhart in *Law of the Plainsman.*[14]

The rider's journey, as is revealed later in the story, comprises the hero's brief but regular returns to his native roots to spiritually regenerate. He travels on horseback to some distant cliff-side dwellings where Naiche (Chief George Clutesi), a Navajo medicine man, lives.[15] Parker goes to visit Naiche for advice and spiritual guidance. He calls Naiche "grandfather," but this does not mean that the old man is Parker's literal grandfather. "Grandfather" is a traditional term of respect for an elder; it is also a name for peyote, since Navajos believe a specific deity resides in the peyote button.[16] Parker is frequently pursued by his young nephew, Half Cub, who tells him that Grandfather wants to see him. It is significant here that Parker has family members. Few if any of the American Indian characters featured in network programs until this point have had family members, isolating them and making their assimilation almost total: they have abandoned their native way of life, including their families, in favor of living among whites; this is most obvious in the cases of Tonto (in *The Lone Ranger*) and Sam Buckhart (in *Law of the Plainsman*). Parker is not nearly as isolated (socially) as these other characters.

When Parker has a life-altering decision to make, he goes to visit Naiche; the two sit in a cliff-side dwelling meditating by candlelight and consume a liquid concoction, probably peyote tea (figure 7.3). In this scene, Naiche's face is lit; he is closer to the camera and on the left, indicating his superiority. Naiche represents authority and Navajo tradition; indeed, it is debatable as to whether the character actually exists or is an allegorical representation of Navajo spirituality, like the peyote itself. Parker follows Naiche's motions, his face obscured by the ladle. Parker is hidden; his emotions are unknown, even to himself. Despite his commitment to Navajo spiritual pursuits, Parker—much like Tonto and Sam Buckhart—has acquired an overriding admira-

Figure 7.3. Naiche and Parker in cave, drinking what is most likely peyote tea.

tion for Anglo-Saxon jurisprudence. He has spent a great deal of time and effort learning the white man's laws; yet, as we shall see, he is relentlessly drawn back to his Navajo roots and his native customs. He is conflicted. The story thus revolves around Parker's "double consciousness," as DuBois called the internal struggle of the hybrid:[17]

> **Parker:** I don't know who I am. I don't know *what* I am.
>
> **Naiche:** You are caught between the earth and the sky.
>
> **Parker:** I'm at war with myself, Naiche.

This same lack of identity was noted by Fanon in 1952: "colonialism forces the people it dominates to ask themselves the question constantly: 'In reality, who am I?'"[18] Parker's being trapped "between earth and sky" is apparent in the contrast between the introduction, in which he is seen in the open and framed against the sky, and his now being imprisoned in this tiny, dark cave. We also see that Parker is apparently not a Christian, at least not in any Anglo-American sense, and this is made even more clear later, when Parker and city attorney Samantha Lowell (Linda Evans) discuss each other's respective deities.[19]

THE PILOT: A MADE-FOR-TV MOVIE

In contrast to both *Law of the Plainsman* (see chapter 4) and *Hawk* (chapter 6), Parker is not the only American Indian character. Besides Parker and Naiche, the pilot contains four other presumably Navajo characters: Diane Little Eagle (Maria Elena Cordero), a community activist/organizer; Little Eagle's friend Peter Two Bears (Joe Renteria); Indian Johnny (Jay Varela), a money-grubbing "Uncle Tom-Tom" (American Indian slang equivalent of Uncle Tom); and Parker's nephew, Half Cub (Jed Horner Jr., replaced in the weekly program by John Tenorio Jr.), who likes to sneak up on Parker. It is noteworthy that nearly all the actors who play American Indian roles have Spanish surnames, again indicating the interchangeablity of Mexicans and Indians in the public consciousness.

Parker is more dedicated to his ideals of social justice than simply enforcing the law and desperately hopes to reconcile these two poles. While returning from a spiritual sojourn, he comes across three white men stealing a steer on Navajo property and intervenes. He is not in uniform; the men do not realize he is a police officer, and he does not identify. Perhaps he wants to see how they will act if he poses as an ordinary citizen or as a simple Navajo; like Clark Kent, he is in his secret-identity mode. He has a secret identity as a hypermasculine being into which he can transform himself by way of some ritual (or emotional) trigger. Not realizing who he is, the men tell him to mind his own business, but he presses on. When they attack him, he easily beats up all three, ties them up, puts them in the bed of his pickup truck, and takes them off to jail. However, when he later discovers these are out-of-work migrants with families to support, he lets them go. In doing so, he demonstrates his fair-mindedness, compassion, and identification with the poor and with outsiders like himself. He is a Christ-like figure. Like the New Testament Jesus, Parker is anti-authoritarian, uncontrollable, unpredictable, and clearly on the side of the poor and dispossessed. Parker's boss, Sheriff Sam Jericho (Arthur Kennedy), is not pleased with Parker's independent spirit and obedience to a higher law, yet Parker is one of his best men. Jericho, a paunchy, middle-aged white man, is somewhat Indianized himself; he wears a Thunderbird (a Navajo religious icon) bolo instead of a necktie and has a Navajo blanket hanging on the wall of his office. The concept of Indianness here seems to suggest *authenticity* and extends to white men (i.e., they are partly Indianized) if they exhibit a sense of social justice rather than strict adherence to hierarchy. "Whiteness" applies to people who are blindly authoritarian or greedy, the latter of which includes Indian Johnny, a Navajo.

The plot of the made-for-TV movie concerns white land developers coming to the fictional town of Concord, New Mexico, where they can purchase land cheaply. They are shifty-eyed men in business suits, common villains in the post–hippie era of the 1970s. There is a city/rural dichotomy at work here with "city slickers" seen as untrustworthy, avaricious, menacing.

The developers are shot from low angles to make them appear more menacing, and the lines of the ceiling point directly to their heads (figure 7.4): as extremely "white" (i.e., materialistic) men, they have brains but no hearts; their overreliance on their brains results only in deviousness (Parker, on the other hand, relies mostly on his heart). As the two men stroll around the area in an extreme long shot, they evince a proprietary, sexualized gaze at the land: "Look at that land—water, rail nearby—and most important of all, cheap." Yet unlike the beautiful hills in the opening sequence, the land these developers are looking at is barren desert. Nonetheless, there is such a demand they can pretty much build anywhere and be assured of a bankable market. This is an allusion to the ideal of perpetual growth or what is often termed "progress." However, from the Navajos' point of view—except Indian Johnny's—such progress has consumed the Natives' land for centuries, and the process has not abated. There is no place safe from encroachment, not even the desert. The text seems sympathetic to the Navajo view: American Indians had already been consigned to barren lands nobody else wanted, and now greedy white men are coming to take even those.

In an interesting twist, the Navajo characters, led by Little Eagle and Two Bears, are not fighting to save a burial mound or some other sacred Navajo site; they are fighting to save a Catholic mission left over from the era when

Figure 7.4. Shifty-eyed real-estate developers in business suits shot from low angle to make them appear more powerful or threatening.

New Mexico was simply Mexico. This suggests the Navajos' connection to the past, yet it is a past that is not their own; again this demonstrates the proclivity to conflate American Indians with Mexicans. The town itself is a hybrid, comprising elements of Navajo, Spanish, Mexican, and Anglo-American cultures. The Navajo activists enlist the support of roughly half the townsfolk, mostly everyday Anglo-Americans who do not want to see bulldozers tearing down local landmarks. However, the other half, mostly impoverished or unemployed, including Indian Johnny, desperately want the influx of tourism and jobs.

The issue of social class looms large but is never directly addressed. Parker is a modernist Navajo who has wanted to get off the reservation since he was young. He was one of the lucky ones: unemployment on Indian reservations runs as high as 69 percent.[20] He has escaped the reservation and now has a steady, respectable job as a sheriff's deputy, but he is also trying to hold on to his Navajo roots. In contrast stands Indian Johnny, whose mixed-race outfit combines a shiny, western-style cowboy shirt, black cowboy hat with a Navajo hatband, and ostentatious Navajo jewelry. Johnny is an ultra-modernist—to the point of seeming like a sellout. He is what some Indians colloquially call an Uncle Tom-Tom or an apple (i.e., red on the outside, white on the inside).

Indian Johnny's Navajo identity seems to be a gimmick that gets him attention. He is happy to pander to white people's stereotypes when it benefits him. He even mocks the Navajos' religion: at one point he sarcastically remarks to Parker, "The Great Spirit told me to expect a tourist stampede." Demonstrating that he is mainly concerned with money, he adds, "Nakia, you'd better ask that Great Spirit of yours how you can get in on the action," knowing that "true" Navajos tend not to be materialistic. In this series, materialism and greed connotes whiteness. If Parker is a hybrid of Navajo and white, Johnny is almost entirely white—or, in Indian slang, an apple: red on the outside, white on the inside. Indian Johnny finally pleads with Parker: "Nakia, we're brothers. Give me a break, will ya? You could change all our lives." Johnny is asking Nakia to look the other way as the developers take over their town.

Magnifying Parker's identity crisis, Nakia encounters two potential love interests: Diane Little Eagle, a Navajo community organizer and activist, and the blond, blue-eyed Samantha Lowell (Linda Evans), a city attorney. His relationship to Little Eagle is not clearly revealed; she might be a girlfriend or a close relative. Little Eagle, in keeping with her civic activism, is loud, aggressive, often angry in public; in private, she is demanding and puts pressure on Parker to be faithful to his people and their traditions. Parker and Little Eagle are alienated despite the fact that they have known each other since childhood.

Their alienation is demonstrated in a scene they share. The two characters keep their distance. Their dialogue is mostly seen in separate shots/reaction shots; they are rarely shown in the same shot, and when they are, Little Eagle's face is obscured by her thick, dark hair. Moreover, they have a huge barrier enveloping them (figure 7.5). Little Eagle expresses her disappointment in Parker, an opinion that weighs upon him heavily. She is framed against the sky with clouds for a halo, suggesting her own connection to the heavens and her spiritual authority (figure 7.5). Her dark hair contrasts with the sky. Her facial expression is severe, earnest. Her authority, however, is undermined by her position in the frame: in most shot/reverse shot sequences, it is the person on the left who initiates questions and the person on the right who reacts. She calls him back to Indianness, a longing voice programs like *The Lone Ranger*, *Plainsman*, and *Hawk* lacked. Although not as gross a sellout as Indian Johnny, Parker already feels like a fink. Little Eagle wants him to stand up for the ordinary people, not the rich developers. She confronts him:

Little Eagle: You've changed, Nakia.

Figure 7.5. Diane Little Eagle framed against the sky, dark hair accented by the contrast.

Parker: Life is change.

Little Eagle: When do you stop letting that badge cripple you?

Issues of masculinity abound. It is fitting that the male-identified world is represented by Parker, who is not entirely firm in his commitment to hypermasculine (western/cowboy) culture. "Male protagonists vary in the exact amount of masculinity they possess," notes Richard Kaplan.[21] The female entity is represented by Little Eagle, who stands for Navajo culture. Western culture is often hypermasculinized, while American Indian culture is feminized—despite the fact that Indian males, as individuals, are often hypermasculinized. Parker has a foot in both camps; his developing hypermasculinity (as a police officer) is being reined in by Little Eagle, who would like him to be more sensitive to human issues instead of an abstract code of laws; her entreaties have a clear effect.

Yet western culture is not entirely void of sensitivity. In the scene immediately following his confrontation with Little Eagle, Parker encounters fair-haired city attorney Lowell, who is quiet and subdued. Her hair color and texture contrasts markedly with Little Eagle's in the previous scene (figure 7.6). Also in contrast to Little Eagle, who chastises Parker, Lowell is receptive; more importantly, she does not make demands.

They are shown strolling together in a slow, lengthy two-shot. Both smile frequently. An opportunity arises for Parker to demonstrate his sensitive (i.e., feminine) side by playing with a tow-headed child, an act that impresses Lowell. Parker has a moderate, tempered style of masculinity. Martin Pumphrey explains, "[Western] heroes must be both dominant and deferential, gentle and violent, self-contained and sensitive, practical and idealist, individualist and conformist, rational and intuitive, peace-loving and ready to fight. . . . Quite simply, the hero's masculine toughness must be partially feminized."[22] Parker puts his arm around her as they exit the scene. At this point both are framed against the sky. In contrast to Little Eagle, who is assertive, almost aggressive (coded as mannish), Lowell is patient and relatively submissive (coded as feminine). Ironically, the Anglo-American woman is clearly the more feminine. These two women personify very different—clashing—civilizations, and it appears Parker at this stage is in the process of choosing one or the other. Choosing which civilization he wants to "marry" seems to be the heart of Parker's dilemma—as well as the theme of the

Figure 7.6. Lowell, too, is framed against the sky.

movie. At this point, Parker appears more comfortable with the Anglo-Saxon Lowell than with one of his own.

The cultural contradictions Parker faces—as well as his own increasingly painful cognitive dissonance—continue to build. He can no longer ignore the town leaders' hypocrisy—nor his own complicity in it—after the developers obtain an injunction charging Little Eagle's group with inciting violence; in fact, the violence has been instigated by the developers' own hired thugs. The film's writers demonstrate that they understand the dynamics of disowning projection (as a form of denial) as Parker and Little Eagle share Navajo wisdom. The two Navajos bond over this bit of wisdom. They note that guilty people often project their own malevolent intentions onto their victims:

> **Parker:** The bear steals the honey from the bee and complains to the Great Spirit because the bee stings him.
>
> **Little Eagle:** The hunter says the eagle is a vicious bird because it attacks him when he steals its eggs.

This brings the two together again. Parker realizes he and Little Eagle share a bond that cannot be broken. He caresses Little Eagle's thick, coarse hair, a signifier of her Indianness, demonstrating that he is becoming more accepting of her and her people, who are also his own.

In addition to racial issues, this movie deals with class divisions and conflicts. Whenever Parker is off duty, he wears denim (shirt and jeans), which emphasizes his working-class roots. Parker is fiercely dedicated to democratic ideals of government by the people, and this meshes with his belief in the law. It also meshes with the grassroots resistance movement Little Eagle has established. Thus the viewer is invited to conflate indigenous culture with a populist, grassroots type of democracy.[23] Parker becomes a sort of Robin Hood figure working to protect the "little people"—not just Navajos, but poor whites as well—from plutocrats who want to bend the law to their benefit at the expense of the community. This story represents a fairly typical conflict that goes on in rural (and colonized) areas when progress arrives.[24] At the same time it incorporates American Indians and their religious practices into the narrative, making the loss of the land even more poignant. The bulldozers become a symbol of both progress and devastation, depending on whose point of view is considered.

Because of his position, Parker is effectively immobilized; at least he seems to think so. While making his living capturing criminals, Parker ironically finds himself trapped. Even worse, he is starting to lose his respect for the law. "Sometimes the law gets in the way of justice," he says. This is a cliché of the sort that might be used by any number of vigilante or rogue-police figures, including the Lone Ranger, Kojak, Billy Jack, Dirty Harry, Baretta, and so on. After he realizes the legal system favors the developers, his badge becomes a symbol of his abandonment of his people, of the democratic process, of his inability to do the right thing. "It's not a badge anymore," he tells Jericho. "It's a pair of handcuffs around my wrists." Parker never considers becoming a vigilante or a rogue cop, however. After Little Eagle and her community organizers lose their case in court, Parker abandons neutrality:

> **Parker:** We must continue this fight. There are many ways to win.
>
> **Sheriff Jericho:** (jaw agape in confusion) We wear badges, Nakia—we don't take sides.

Parker yanks off his badge and hands it to Jericho. He has finally made his decision.

Parker goes home to brood. The protest turns into a confrontation. The protestors are a multiethnic group with a large segment of swarthy types including Navajos and Mexican Americans. There are no African Americans, however. The counterprotestors, led by the developers' thugs, are entirely white. A riot ensues. Seeing the riot, Parker realizes he can do more good as a policeman than as a private citizen. His respect for the impartiality of the law is suddenly, inexplicably restored. He runs to the scene of the melee to help his police brethren restore order. He decides he is a police officer at heart. Jericho is eager to welcome him back. Parker is recuperated into white patriarchy, but only as a junior partner. He is now happy to accept this position.

Parker is visually humbled before the camera (figure 7.7). He is situated to the bottom right of the frame, the least-powerful sector. He almost appears to be kissing Jericho's feet (figure 7.7).

Why Parker arrives at this last-minute change of heart is not addressed. This is uncharacteristic in that during most of the movie the viewer is privy to all the details and difficulties of Parker's ongoing *angst*, yet now he arrives at a decision instantly and painlessly. The riot serves as deus ex machina that magically dissolves his dilemma and rights the story. The riot externalizes Parker's mental state and his own inner battle. More likely, however, this development has an institutional basis: Parker *must* remain a policeman if the movie is to become a weekly program. If he had quit for good, there would be no series, at least not with Parker as a policeman: a series about a Navajo janitor might not be as interesting or exciting. Media critic Robert Cirino explains, "[I]f a writer wants to communicate a political message by having the main character quit his job in protest, he or she is at a disadvantage because weekly television programs demand that the hero . . . maintain a

Figure 7.7. Parker prostrates himself at Jericho's feet, Jericho in a godlike pose.

certain image or outlook and continue to overcome difficulties *without changing the system*."[25]

WEEKLY SERIES

The made-for-television movie was adapted into a weekly series for ABC, where it ran from September 21 to December 28, 1974 (only thirteen episodes). The weekly version differed greatly from the movie: it became more and more like Gerber's own *Cade's County* and less like *Billy Jack*.

Its new opening sequence now consists of still photos that showcase each of the main characters in bright, loud, contemporary graphics using a pop-art, Andy Warhol–influenced style, signifying that the program is young, hip (see figure 7.8). These are alternated with still shots of unspecified Indians doing a pan-Indian pow-wow dance (figure 7.8) as well as a few shots fetishizing the land. The contrast alludes to Parker's conflict between modernity and tradition. There is no voiceover, just western-sounding music with minor-modal overtones and simplistic harmonies voiced in fourths, which suggest

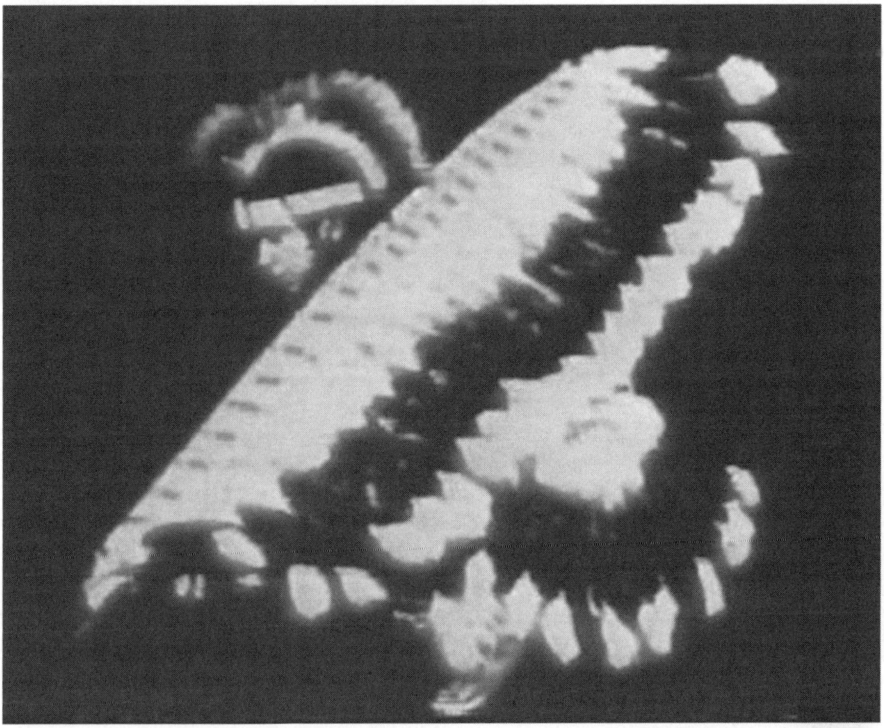

Figure 7.8. Pan-Indian figure dancing, depicted in an exaggerated high-contrast design much like Warhol's.

Indianness. There are occasional helicopter shots, and there is less handheld work than in most modern police or action dramas (*Hawk*, for example, discussed in chapter 6, frequently relied on a handheld camera).

There are lots of close-ups, usually with Parker on the left side of the screen, indicating his power. Again, he is often framed against the sky in godlike poses, sometimes with clouds as a halo (figure 7.9). He looks to the sky as if waiting for a message. As in the pilot, Parker's outsider status renders him hip and sensitive. Like John Hawk, who was a mediator between generations as well as between urban hipsters and the Establishment, Parker is a mediator between Navajos, poor whites, local hippie types (all conflated) and the Establishment. Parker's superior tracking skills enable him to find criminals easily, and he is very much at home in the wild. He is associated with wilderness. Parker's native survival skills save his life when he is stranded in the desert, a situation in which an ordinary Euro-American would perish (episode 6, "A Beginning in the Wilderness").

Figure 7.9. Parker looks to the sky for answers.

There is not much direct reference in the series to Parker's Indianness except for an occasional, good-natured joke. However, in one episode, Sheriff Jericho enthusiastically says he intends to "hire more Indians" (episode 12, "Pete"). This statement, almost a sales pitch for American Indians, might represent the model-minority trope (discussed in chapter 4) that is conjured when race and class relations between middle-income whites and urban blacks become tense, and this was the case after a federal court's 1971 busing ruling. Other than this episode, the issue of Indianness fades into the background in the weekly series, whereas it was the primary focus of the pilot. Veteran actor Michael Ansara appears in what may be an intertextual nod to *Law of the Plainsman* and *Broken Arrow* (episode 9, "The Dream"); however, a viewer would need to be familiar with these 1950s westerns to understand the reference (on the other hand, it is possible that Ansara was merely typecast due to his swarthy look).[26] Ansara plays Navajo physician Dr. Howard Gray Hawk, echoing his role as the highly educated Apache, Sam Buckhart, in *Law of the Plainsman* (chapter 4). Gray Hawk is a symbol of "bootstrapping," living proof that any American, no matter what his or her ethnic background, can be upwardly mobile (in class terms) as long as he or she is willing to work hard enough (i.e., "pull himself up by the bootstraps"). This reinforces the model-minority trope: the implication being that if other minorities, specifically African Americans, fail to "make it," it can only be a result of their lack of work ethic. Like Parker, Gray Hawk is caught between two worlds. Even though he is a certified medical doctor, a man of science, he still believes in the Navajo traditions, and he dresses in denim. He tells the sheriff, "I'm still an Indian, Sam; I always will be."[27]

Episode 10 ("Roots of Anger") deals indirectly with Navajo views on wildlife. As shown in the pilot, Parker and his Navajo family members revere all creatures. "Even the snake is our brother," Naiche says. The original script for episode 10 features the following remarkable exchange between Parker and his young Navajo nephew Half Cub, who had been playfully trying to sneak up on Parker. Parker, using his intuitive tracking and listening skills, tells Half Cub he knew he would be coming from a certain direction because earlier he had noticed an ant hill. The original script reads as follows:

> **Nakia:** I didn't really hear you detour around the ants' nest. I guessed that part.
>
> **Half Cub:** How come?
>
> **Nakia:** I knew no Indian would ever step on an ant if he could help it.
>
> **Half Cub:** Maybe that's the trouble with Indians—we won't step on ants.
>
> **Nakia:** Maybe that's the trouble with Anglos—they will.[28]

However, by the time this dialogue appeared on air, the reference to "Anglos" had been softened:

> **Nakia:** No Indian would ever step on an ant if he could help it.
>
> **Half Cub:** Maybe that's the trouble with us Indians—we won't step on ants.
>
> **Ben:** (a third Navajo character) Maybe that's the trouble with *everybody else*—they do step on ants. (emphasis added)

In the original script, Nakia obliquely accuses Anglos of not respecting wildlife nor animals' rights. But ants might also be a metaphorical reference to *people* deemed insignificant by the dominant group. His comment implies that Anglos are generally quick to "step on" anyone who gets in their path—such as indigenous peoples. This could be construed as emblematic of a colonial mind-set. The quasi-official stance is that the United States does not colonize other countries nor does it step on other peoples. The Navajos here could be implying that Anglo culture is intrinsically colonialist. Columbia's production executives may have found this potentially insulting to ABC's predominantly Anglo viewership, or it might have proven an embarrassing item critics of the United States might have seized upon; after all, the Cold War was still ongoing in 1974. It does seem as if the writers are more in touch with—and sympathetic to—Native views than the program's producers and network executives might appreciate.

Even though the *pilot* depicted Parker as a conflicted individual caught between two worlds, most episodes in the weekly series feature him as an ordinary police officer—who happens to be a Navajo—simply doing his job. In episode 7 ("The Driver") the villains again are white city slickers in business suits. Again, they are real-estate investors, looking to buy cheap land. The episode's writers seem aware of the colonizing aspects involved in the exploitation of Native lands—the villains have come specifically to exploit the area's uranium reserves, which was an important aspect of the "termination" of American Indian treaties during the Cold War (discussed in detail in chapter 3 on *Broken Arrow*) and continued until 1970 when termination was renounced by President Nixon.[29] The writers also allude to the high unemployment rate on the Navajo reservation.[30] A young man named Brennan (Geoffrey Deuel) finds work as a driver for two white businessmen. Brennan is an Indianized Euro-American who has married a Navajo woman (played by Victoria Racimo, another Spanish surname) and has moved in with her family on the reservation. Brennan now lives by himself in a trailer near town because there is no work on or near the reservation. He has left his wife and infant son on the reservation (Ben, a Navajo elder, perhaps the infant's literal grandfather, calls the boy "little warrior").[31] Brennan serves as—or reminds Parker of—a younger version of himself: Parker, too, might once have willingly or unwillingly turned to crime as a means of escaping unemployment and poverty. Brennan is saving the money he is earning to make a down payment on a house in town for his family. He is moving along the path of upward mobility, but the path is too easy—something must be wrong, as Parker correctly intuits: the men in suits are paying Brennan extremely well but ultimately intend to eliminate him. Suits and ties in this program signify greed and rapacity. Brennan, like Nakia, dresses in working-class denim.

Just as Nakia is Anglicized, white men here are Indianized, to varying degrees. The red Navajo blanket on the wall of Jericho's office (figure 7.10) is like a flag, and it appears as if touching Sheriff Jericho; the diamond-shaped symbol points directly at his head. This suggests that Navajo ideas have entered his mind. It serves a similar purpose for Parker, an identifier of his Navajo nationality (figure 7.10). Both Parker and his fellow officer are

Figure 7.10. Shadows from Venetian blinds envelop Parker and another deputy; Navajo diamond design "touches" Jericho's head.

shaded by Venetian blinds. This 1940s film noir cliché suggests that the characters are in doubt about something or perhaps being willfully ignorant, refusing to "see."[32] The shadows here are like rays emanating from the blanket, which in turn points directly to Jericho, who begins to "see the light." Later, in a reference to the character's messianic heroism, Parker, like Superman, leaps from the sky (actually from a helicopter) to pounce on a hired assassin and put him in handcuffs.

Episode 11 ("A Matter of Choice") echoes the theme of the pilot. Parker, however, plays a sideline role in this while another officer becomes emotionally involved in a case against a drug manufacturer who sells black-market pills to teenagers. Again the villain is a white man in suit and tie. White businessmen in suits and ties are common villains in U.S. films, especially those with a populist appeal. "Popular culture possesses a reservoir of anti-capitalist meanings that are continually drawn upon to animate the mass narratives of our industrial culture."[33] However, depicting investors as villains does not represent an attack on capital per se, as Susan Hayward points out; these are, specifically, "crooked" capitalists: "In some film narratives, capital is corrupt, so [it] must be resisted. However, since that representation is often caricatural, capital as a bad thing is not being targeted but rather the abuse of capital power, for which an *individual* will be punished."[34] Churchill elaborates,

> Euroamerican bad guys are *always* individuals, "criminals," who have deviated from the rules. . . . They are never representatives of the governmental [or major] corporate entities. . . . Both conservative and liberal Euro-Americans are delighted to learn that the source of daily agony for oppressed people of color lies not in the structure of white supremicism but in deviation from its orders and values.[35]

Slavoj Zizek remarks that "all the talk" in contemporary society about ethnic minorities and cultural change serves as a distraction; ensuring that "*something*—what really matters—will not be disturbed." This "something" is "the inexorable logic of capital."[36] Again, this is reminiscent of Lasswell's 1945 recommendation to Benton: "play up" the aspects of progress in order to "to present a full and truthful picture of life in the United States." Yet at this time little of substance was being done to ameliorate systemic racism in the United States. Basically Lasswell is recommending that the state should intervene for the sake of appearances while continuing to allow the marketplace to make its own decisions.[37]

The subtext of the story at this point is that the state should and will intervene to protect ordinary people from abuses of power by out-of-control capitalists. This story is trying to satisfy both sides, reassuring the "plain folks" that the government will protect them while assuring capitalists it will not intrude on their prerogative to acquire unrestricted profits. Parker himself is one of the plain folks; he comes from the poorest of backgrounds, the reservation. However, Sheriff Jericho insists the law must not take sides. Like Parker, Officer Barrows (in a different episode) quits the police force, but unlike Parker, he decides to take the law into his own hands by becoming a rogue cop, a vigilante. Once again Jericho counsels deference to the system: "Let the law take its course." The sheriff's message is especially significant compared to theme of *The Lone Ranger* (discussed in chapter 2). Even though the Ranger constantly mouthed platitudes about the sanctity of the law, he was all too eager to violate it by faking his own death and becoming a roving vigilante. The subtext of *Nakia*, on the other hand, is "*don't* be a Lone Ranger—trust the system."

HISTORICAL CONTEXTS

After *Nakia* was made into a weekly series, the program's ideology had little connection with race, representation, or American Indian issues. Unlike the pilot, in which Parker's racial identity and loyalty is the key issue, in the weekly program race is downplayed. It should also be noted that there are no African Americans in this weekly series, not even as extras. The series, for the most part, posits the orthodox cop show ideology that had been a staple on U.S. television: police officers nearly always administer justice fairly and impartially—if not always dispassionately—therefore, the public should trust them. Michael Parenti notes that this law-and-order ideology serves to inspire public confidence in the police and in government: "Media dramas teach us that . . . the police and everyone else [in government] should be given a freer hand in combating the criminal element, using generous applications of violence without too much attention to constitutional rights."[38]

One important factor in this was the ethics code of the National Association of Broadcasters (NAB). This code insisted that respect for the law was sacrosanct. Many television executives interpreted the code to mean that the concept of law and order was never to be questioned; viewers were expected to take the pursuit of justice as an article of faith. The NAB code goes back to the Korean War, during which President Truman proclaimed a state of emergency that allowed his administration broad powers of censorship.[39] "[Television] industry leaders pledged their willingness to cooperate with the government in every feasible way."[40] This was still the case in 1974; in fact, Truman's 1950 state of emergency had still not officially been lifted.[41]

Nakia's police-drama ideology is better understood when considered in its specific historical context. Nixon and Agnew had been elected in 1968 on a platform that incessantly invoked the theme of "law and order" and were reelected in 1972 in a landslide victory. Nixon correctly concluded that public assurances that his administration would do everything in its power to restore normalcy were what the silent majority wanted to hear. However, the silent-majority voters Nixon courted were exactly the middle-aged demographic the networks sought to jettison in favor of the nineteen- to thirty-four segment (see discussion in chapter 6). ABC needed this younger audience yet could not afford to alienate older viewers—or the Nixon administration; Nixon was still president in April (he resigned in August) when *Nakia* made its debut. So the question became how to offer younger viewers relevance and rebellion and at the same time retain older, conservative viewers. The solution was to *co-opt* youth rebellion by offering a titillating taste of revolt that reversed itself at the last minute, ultimately justifying and *strengthening* the status quo.

Alan Sinfield discusses this misleading strategy he calls the *entrapment* model: the reader is lured into what he or she thinks is a subversive or dissident text, only to discover at the last minute it becomes a conservative text in which the status quo is affirmed. Sinfield asserts, citing Raymond Williams, that this strategy goes back as far as Shakespeare.[42] Whether or not it is true of Shakespeare, it is, however, demonstrably in certain crime series, especially those involving minority police officers like *Nakia* or *Hawk*. This is unusual and attention-getting because many viewers might expect minorities to oppose rather than support the status quo. Perhaps the best example of this was Aaron Spelling's 1968 series *The Mod Squad*, featuring a young trio consisting of a hip white guy; an idealistic, white hippie female; and a black-nationalist radical, who worked to support the Establishment.

Nakia, the made-for-TV movie that aired in April, offered a story designed to appeal to rebellious youth. Parker's "double consciousness" and struggle with identity—the movie's very focal point—are largely eliminated in the weekly series to make room for Nixonian law-and-order themes.[43] Though it had begun as a complex, ambivalent portrayal of a Navajo lawman struggling with issues of identity, in the end it became a rather ordinary cop show with a rural, western setting and a character who happened to be a Navajo. Although it occasionally did deal with questions of racial representation and social justice, these aspects were minimized as the series conformed to a traditional cop-show formula, albeit with a slight Gerberesque twist.

Like *Law of the Plainsman*, *Nakia* represented the failure of the Indian Regulator to draw a large enough audience to sustain a network series for more than one season. *The Lone Ranger* lasted eight years, probably because it featured a *white* savior, as did *Broken Arrow*, which ran for four. In 1959–1960 the U.S. public was apparently not in the mood to accept an American Indian authority figure in *Law of the Plainsman*, and apparently neither was it in a mood to accept one in 1966 with *Hawk* nor in 1974 with *Nakia*. The Indian lawman remained a failed experiment until *Walker, Texas Ranger* in 1993. Perhaps the predominantly white viewing audience was uncomfortable seeing a nonwhite savior figure.

The figure's Indianness needed to be downplayed and *yet remain accessible* when required to give the hero a "testosterone boost" in dealing with an extraordinarily tough case or tough criminal—in other words, when hypermasculinity was required. American Indians in television and film texts are characterized as aggressive (hypermasculine) and physical, yet paradoxically they are also feminized (see chapters 3 and 4). The concept of American Indian masculinity seems to be extremely fluid and contradictory: it appears to depend on whichever aspect is required to make the white man appear more impressive; in other words, these texts want it both ways. The solution that evolved was to synthesize the Indian and white characters: the protagonist became a half-breed. The white savior could then incorporate the hypermasculine tropes of the Indian without the feminine ones. The half-breed would be a (mostly) white man with a secret Indian identity or "alter ego," like Superman or the Hulk.[44]

NOTES

1. Edward Said, *Culture and Imperialism* (New York: Vintage Books, 1993), 29.
2. Kirby Farrell, "The Berserk Style in American Culture." *Cultural Critique* 46, Autumn 2000, 201.
3. Herman Gray, "The Politics of Representation in Network Television." In Horace Newcomb, ed. *Television: The Critical View*, 6th ed. (Oxford: Oxford University Press, 2000), 282–302.
4. Vine Deloria Jr., *Custer Died for Your Sins: An Indian Manifesto* (Norman: University of Oklahoma Press, 1988), 172–73.
5. Erik Barnouw, *Tube of Plenty: The Evolution of American Television* (Oxford: Oxford University Press, 1990), 430.
6. Also in 1974, ABC presented *Kodiak*, featuring a white Alaska state patrolman, Cal "Kodiak" McKay, played by Clint Walker, who had played the Indianized Cheyenne Bodie on *Cheyenne* (1955–1963). *Kodiak* included a Native (Inuit) character in the role of ethnic sidekick: Abraham Lincoln Imhook (Abner Biberman) served as friend and confidante to the white lawman. *Kodiak* lasted only four episodes, becoming one of the shortest-lived series ever.
7. Sam Pack, "Constructing 'the Navajo': Visual and Literary Representation from Inside and Out." *Wicazo Sa Review* 15(1), Spring 2000, 141.
8. Another Elvis character, Tulsa McLean, an army enlisted man in *G.I. Blues,* was part Indian as well—and also named Joe.
9. Tony Hillerman, *The Blessing Way* (New York: HarperTorch, 1990).
10. "*Nakia*: Synopsis." MSN Movies. http://movies.msn.com/movies/movie-synopsis/nakia/.
11. *Billy Jack* was in turn similar to *Navajo Joe* (MGM, 1966), in which Burt Reynolds starred. Laurence Laurent, "Law 'n' Order Still Dominates New Schedule." *Washington Post*, July 28, 1974, TV-7; see chapter 6 for details).
12. Will Wright, *Six Guns and Society* (Berkeley: University of California Press, 1977), 58, 82. Indeed, the western and the crime drama share a similar genealogy as well as ideology; I shall discuss these confluences in chapter 9.
13. This is an Oglala prayer, not Navajo. "Prayer to the Great Spirit." Peace Winds. http://www.peacewinds.com/main/native/prayer.html. Hollywood writers often conflate a given American Indian group's characteristics with other groups' traditions, as if they were interchangeable. A single "great spirit" is not a Navajo concept. It is "pan-Indian with Christian undertones." Leighton C. Peterson, e-mail to author, June 12, 2013. *Nakia*'s writers do get it right, however, when Parker mentions the Spider Woman, the Navajos' most sacred deity. "The Spider Women." *Time*, September 11, 1972. http://www.time.com/time/magazine/article/0,9171,906347,00.html.
14. Navajo Joe, too, in Sergio Corbucci's 1966 eponymous spaghetti western, was a savior figure.
15. This is another instance of Hollywood writers treating native peoples and their traditions interchangeably: these cliff-side dwellings were most likely built by the ancient Anasazi. Due to their religious beliefs it is unlikely Navajos would take up residence in them. Leighton C. Peterson, e-mail to author, June 12, 2013.
16. Richard Erdoes and Alfonso Ortiz, *American Indian Myths and Legends* (New York: Random House, 1985), 68.
17. W. E. B. DuBois. *The Souls of Black Folk* (Rockville, MD: Arc Manor, 2008), 12.
18. Frantz Fanon, *Black Skin, White Mask* (New York: Grove Press, 2008), 250.
19. Certain Native American groups such as Cherokees have managed to reconcile Christianity with their native religions (Richard Allen, e-mail to author, 2010).
20. Tom Rodgers, "Native American Poverty: A Challenge Too Often Ignored." *Spotlight on Poverty*, September 15, 2009. http://www.spotlightonpoverty.org/ExclusiveCommentary.aspx?id=0fe5c04e-fdbf-4718-980c-0373ba823da7.
21. Richard L. Kaplan, "Spider Man in Love: A Psychoanalytic Interpretation." *Journal of Popular Culture* 44(2), April 2011, 294.

22. Martin Pumphrey, "Why Do Cowboys Wear Hats in the Bath? Style Politics for the Older Man." *Critical Quarterly* 31, October 1989, 78–100. Reprinted in Douglas Pye and Ian Cameron, eds. *The Movie Book of the Western* (London: Cassel, 1996), 52.

23. It is true that Native Americans generally have less hierarchical societies than Europeans. In fact, the United States' "Founding Fathers" got many of their ideas about democracy from the Iroquois; Bruce E. Johansen, *Forgotten Founders: How the American Indian Helped Shape Democracy* (Cambridge, MA: Harvard Commons Press, 1987).

24. Said, 1993, 225–26.

25. Robert Cirino, *We're Being More Than Entertained* (Honolulu: Lighthouse Press, 1977), 53; emphasis added.

26. "Michael Ansara—Perennial Indian," *TV Guide*, April 1960, 29.

27. This line may be a sly reference to the 1960 *TV Guide* article cited in note 26.

28. Charles Larson, "'The Non-Person,' Final Draft" (episode script). *Nakia.* David Gerber Productions. June 13, 1974, 33. Ants are considered an incarnation of holy people in Navajo religion; therefore it is crucial for Navajos not to step on them.

29. Many texts instinctively recognize the role of the land's resources and minerals. For example, a greedy rancher who sets out to claim Indian lands rich in silver was the villain in a 1956 feature-film version of *The Lone Ranger*. Chuck Thompson, "Great Movie Indians." *Cowboys & Indians*, June 2011, 67.

30. S. J. Ludescher, "Report: Navajos lag behind US." Gallup (New Mexico) *Independent*, July 19, 2009. http://www.gallupindependent.com/1999-2001/7-18-00.html.

31. Ben appears to be a replacement for Naiche, who does not appear in the weekly series. Little Eagle and Indian Johnny have disappeared as well.

32. Richard Gilmore, "The Dark Sublimity of *Chinatown*." In Mark T. Conard. *The Philosophy of Neo-Noir* (Lexington: University Press of Kentucky, 2007), 122–24.

33. Kaplan, 2011, 304.

34. Susan Hayward, *Cinema Studies: The Key Concepts* (London: Routledge, 2000), 62; emphasis added.

35. Ward Churchill, *Fantasies of the Master Race: Literature, Cinema, and the Colonization of American Indians* (San Francisco: City Lights Books, 1998), 84–85; emphasis in original.

36. Slavoj Zizek, *The Ticklish Subject: The Absent Centre of Political Ontology* (London: Verso, 1999), 354; emphasis in original.

37. This was probably Eisenhower's approach to the 1957 Little Rock crisis as well.

38. Michael Parenti, *Make-Believe Media: The Politics of Entertainment* (New York: St. Martin's Press, 1992), 2.

39. Anthony Leviero, "President Declares a National Emergency." *New York Times*, December 17, 1950, A1. Also see Presidential Proclamation No. 2914, 15 Federal Regulation 9029, 64 Statute A454. December 16, 1950. Reprinted at *US Code Collection.* Cornell University Law School. http://www.law.cornell.edu/uscode/html/uscode50a/usc_sup_05_50notes.html.

40. Nancy Bernhard, *US Television News and Cold-War Propaganda, 1947–1960* (Cambridge: Cambridge University Press, 2003), 97–98. The NAB code was technically voluntary, and only about 65 percent of U.S. broadcasters subscribed to it; it was abandoned in 1982 because sections were seen by the Department of Justice as violations of antitrust laws.

41. Cirino, 1977, 52–53.

42. Alan Sinfield, *Faultlines: Cultural Materialism and the Politics of Dissident Reading* (Berkeley: University of California Press, 1992), 47–51.

43. A great deal of tension had developed during this period between media owners and the Nixon administration. Nixon made a point to harass media owners who criticized him (he had included many media figures in his "enemies list"). The media may behave as Ideological State Apparatuses when their interests *coincide* with those of the state, but U.S. media owners are a powerful, organized group and have been known to counterattack when cornered: for example, it is worth noting which side prevailed in Nixon's battle with the media. CEO William Paley personally approved CBS News' repeated exposés of Nixon's crimes; see William S. Paley, *As It Happened: A Memoir* (New York: Doubleday, 1979), 335–50. This had happened before: In

1954, William Murrow, host of CBS's *See It Now*—with Paley's tacit approval—had focused critical scrutiny on Senator Joseph McCarthy, discrediting the senator and effectively ending his career. Fred Friendly, *The Good Guys, the Bad Guys, and the First Amendment: Free Speech vs. Fairness in Broadcasting* (New York: Random House, 1975), 12–14.

44. This synthesis seems to have first occurred in the film *Hondo* (1953, starring John Wayne; adapted as a television program in 1967), and again in *Billy Jack* (1971), and later in *Rambo: First Blood Part II* (1985), in which John Rambo is revealed to be half Indian. Rambo's Indianness was a device added by *Part II*'s producers and was not part of David Morrell's source novel (Morrell 2009). Marvel Comics' character the Hulk appeared on CBS from 1978 to 1982.

Chapter Eight

The Indianized White Hero

Walker, Texas Ranger *(1993–2001)*

The white man . . . will never let go of the Indian image because he thinks that by some clever manipulation he can achieve an authenticity which can never be his.—Vine Deloria Jr.[1]

The Puritans in Massachusetts defined themselves by comparisons with the Pequots, who were touted as everything the Puritans were not. Yet at the same time the settlers admired them and in some ways wanted to be like them.[2] After the colonists broke from England, however, they needed a new identity. Suddenly it became important for Anglo-Americans to differentiate themselves from "imperialist" Europeans.[3] The answer was the so-called New Man.

For example, the instigators of the Boston Tea Party found it advantageous—for whatever reasons—to dress up as Mohawks.[4] The Revolutionary War–era exploits of Colonel Daniel Boone provided a template for many "Indianized" white heroes who would follow. Several fictional characters with American Indian characteristics, many much like Boone, began appearing in novels, such as Cooper's Nathaniel (Natty) Bumppo. Other Indianized white heroes would follow, including Tom Jeffords in Arnold's *Blood Brother* and, on television, Cheyenne Bodie on *Cheyenne* (ABC-TV, 1955–1963). Coming from the other direction were Indians in various stages of becoming Europeanized, indicating their endorsement of European norms. Like the Indianized white heroes, these figures were cultural hybrids, physically Indian but intellectually and spiritually "apprentice white men." Then there were the "half-breeds" who literally embodied the best of both races: Mingo on *The Adventures of Daniel Boone* (NBC, 1964–1968), Hondo Lane on

Hondo (ABC, 1967), and Cordell Walker of *Walker, Texas Ranger* (CBS, 1993–2001).

This chapter examines the final (or current) stage in the fusion of the white man and the native: the white man who can become native at will. Cordell Walker of *Walker, Texas Ranger* (CBS, 1993–2001) is a half-Cherokee lawman. His "Indianness" is usually kept in the background or rather bubbling under the surface: most of the time he is Anglo-American, but he can become a Cherokee anytime it serves his purposes. Indianness is his secret identity, which emerges whenever superhuman or spiritual qualities are needed—whenever the Euro-American intellect is not enough. Walker can "shape-shift" when the need arises and, unlike Nakia Parker, rarely if ever agonizes over his identity. For Walker, identity is fluid: he can switch back and forth from white to Indian almost at will—all it takes is some ritualistic or emotional trigger.

This chapter also examines issues of masculinity—especially the construct of a specifically "American" masculinity that borrowed freely from ideas about Indian physicality. *Walker* happened to appear at the same time American masculinity and patriarchy faced cultural and political challenges from the women's movement and responded to these challenges. This chapter will also examine some of the reactionary political as well as religious ramifications of the series.

BACKGROUND OF THE SERIES

Walker's star, Chuck Norris, was instrumental in developing the series. Norris brought a good deal of star power and intertextual meaning(s) to the series, which included his background in martial-arts movies. Norris's early career plan had been to become a police officer, and he served a short period in the U.S. Air Force as a military policeman. While stationed in South Korea, he began studying martial arts.[5] After winning several titles, he met Bruce Lee and landed a small role as Lee's nemesis in *Return of the Dragon* (1972). Norris became a cult figure in the hypermasculine action-movie genre, starring in a series of military-related films as Colonel James Braddock, produced by Cannon Films, including *Missing in Action* (1984), in which he operates behind enemy lines in North Vietnam to rescue U.S. prisoners of war, and as Major Scott McCoy, a military commando battling Middle Eastern terrorists in *Delta Force* (1986).[6] Norris told an interviewer that the TV character Cordell Walker was a combination of characters he had played in his movies.[7] In 1983 he played a somewhat similar character in his film *Lone Wolf McQuade*, in which he plays an indefatigable Texas Ranger who counts martial arts among his law-enforcement techniques.[8] Norris is also the author of two books in which he explains his martial-arts philosophy as well as his religious and political beliefs, some of which are manifest in the program.

Walker, Texas Ranger made its debut on CBS in 1993 and enjoyed a remarkably long run, 196 episodes. Cannon Films, with whom Norris had enjoyed a long and fruitful relationship, went bankrupt during *Walker*'s first season; CBS stepped in with funding to keep the series in production. Columbia Pictures Television, the company that produced both *Nakia* and *Hawk*, was awarded the production contract. Norris, who held a good deal of negotiating leverage at this point, was made executive producer and given a profit participation.[9]

Walker made its debut three years after *Dances with Wolves* stirred a renewal of interest in Native Americana. *Walker* is the only long-running network program to feature an Indian (actually, half-Cherokee) lawman in a starring role. Norris himself purports to be half Cherokee—according to his autobiography, both his mother and father are half Cherokee.[10] The Cherokee area in Oklahoma, where Norris is from, happens to be about two hundred miles from Dallas, where *Walker, Texas Ranger* is set. *Walker* included a "full-blood" Cherokee character, Cordell Walker's uncle, Ray Fire Walker, played by Floyd "Red Crow" Westerman, a Lakota actor who had been prominent in *Dances with Wolves*. Uncle Ray, a former U.S. Marine (Westerman himself was a former marine), raised Walker and calls him "Washoe," which means "lone eagle."[11] He lives in Walker's household as a sort of combination ranch hand/manservant. Thus Walker has his own "Indian companion."[12] Walker in fact has *two* "ethnic sidekicks": Like Detective John Hawk, Walker has a black partner, fellow Ranger James Trivette (Clarence Gilyard Jr.), who escaped the ghettoes of Baltimore playing professional football (Gilyard himself played football in college). Trivette is similar to *Hawk*'s African American detective Dan Carter in that he serves as a sort of humorous foil for Walker; however, Trivette is far more assertive than Carter and is given far more screen time.

In 1993, the year *Walker* made its debut, George Gerbner found that "the world of television [was still] frozen in a time warp of obsolete and damaging representations."[13] This was particularly true of *Walker*: the social scenery in Texas seemed not to have changed much since *The Lone Ranger*. There are few minority characters in the program other than Trivette (and some of Trivette's African American friends) and some Mexican criminals. Early episodes inserted ironic intertextual references to *The Lone Ranger*, which gave the program a tone of pastiche. Both *Walker* and *The Lone Ranger* feature Texas Rangers, both are set in Texas, both have a recurring American Indian character who is subservient, both heroes are saintly and refuse to kill. Just as the Ranger rarely shot to kill but merely to injure, Walker aims to disarm his enemies or if possible give them a martial-arts-style beating.[14] As was true of *The Lone Ranger*, *Walker*'s villains are most likely to be white. The villains on *Walker* are often Establishment figures gone bad: in the first four episodes, villains included a mentally ill former CIA operative, a mentally ill former sheriff, a mentally ill parole officer, and a mentally ill Christian-cult leader.[15] In the introductory sequence, there is a brief shot of Walker on horseback racing from left to right at a full gallop, an almost exact duplication of *The Lone Ranger*'s introductory sequence.

Both programs begin with a truck shot of a horse and rider galloping from left to right (figures 8.1 and 8.2). Both characters are framed against the sky, Walker even more so, his black hat contrasting starkly. The Lone Ranger (figure 8.1) is associated with mountains, suggesting strength.[16] The sky is a recurring visual motif in *Walker* (figure 8.2) as it was in *The Lone Ranger* as well as in *Nakia* and in many westerns (see chapter 2 for detailed discussion). The sky, of course, is where the gods live, the implication being that Walker is one of them—if not a god himself, then sent by God. Establishing shots of the Texas landscape immediately situate the program in the tradition of the western.

Figure 8.1. The Lone Ranger and Silver galloping (the rider is Clayton Moore himself).

Figure 8.2. Walker and horse at full gallop (rider is a stunt double).

Like *Nakia* (see chapter 7), Walker could be described as a contemporary—as opposed to quasi-historical—western. However, whereas sheriff's deputy Nakia Parker wrestled with issues of identity and being caught between two worlds, Walker is comfortably ensconced in the white world but can and does call on his Cherokee identity whenever it seems propitious. Walker's identity is rarely at issue.[17] His Indianness is always lurking just under the surface and can be summoned at will, like Clark Kent's Superman (*The Adventures of Superman*, syndicated, ABC, 1952–1958), or triggered by rage, like Dr. David Banner's "alter ego," the Hulk (*The Incredible Hulk*, CBS, NBC, 1977–1982).

NARRATIVE AND VISUAL ANALYSES

References to American Indian concerns are relatively rare in the series but do occur. *Walker*'s first episode, "One Riot, One Ranger," a two-hour, made-for-TV movie that served as the program's pilot, offered a bit of "backstory" about Walker's parents. His father, a Cherokee rodeo rider (a combination cowboy *and* Indian), was beaten to death by racists angered at his being with a white woman. Both parents were murdered. Walker explains, "They started saying all these crude things to my mom, like how could she be with a dirty, rotten Indian and bring a half-breed into this world. My dad was a very proud man, and he confronted these guys."[18] This device seems to have originated in *Detective Comics*, in which Bruce Wayne, Batman's alter ego, swears to avenge his parents' deaths at the hands of robbers.[19] Indeed, Walker is very much a comic-book-type character, as was the Lone Ranger, who also seems to have a lot in common with Batman (or vice versa). Walker is also remarkably similar to a 1980s animated television character named Bravestarr, who was a combination cowboy *and* Indian and served as an intergalactic federal marshal set in the future.[20]

Episode 11 ("The Legend of Running Bear") is devoted to native issues and could have come from the writers of *Nakia*. The plot of this episode concerns two corrupt FBI agents who are working for a mining operator who has discovered uranium on Cherokee land and bribed a county official to have property lines redrawn. The issue of uranium on native land is an important one.[21] These murderous FBI agents aim to eliminate local Cherokees who have uncovered their plot. It is highly irregular for a network-television series to defame officers of the federal government (*Walker* does this elsewhere with a rogue CIA agent). Walker takes pains, however, to explain to his coworkers that "any organization as large as the FBI is bound to have one or two bad apples." There are references to Cherokee spirituality, with Walker and Uncle Ray taking part in a sweat lodge and Cherokee men meeting and smoking tobacco in a tipi. However, sweat lodges are not part of Cherokee culture and Cherokees never lived in tipis.[22]

Although he participates in native rituals when it suits his need to become Indian, Walker seems skeptical of native religion to the point where Uncle Ray deems it necessary to admonish him not to "make fun of the spirits." Walker visits his Cherokee cousin David Little Eagle, who lives on a nearby Cherokee reservation and is in some sort of trouble.[23] Walker's presence on the reservation seems unremarkable; he is treated like an ordinary white man who happens to have a friend on the reservation and is largely ignored. On the other hand, several Cherokee toughs give cousin Little Eagle a hard time about being a "traitor" because he has left the reservation to become a "city boy" and study medicine (much as Nakia Parker left the Navajo reservation to become a police officer). This paints these Cherokee youths as anti-intellectual and disdainful of education. Oddly, however, none directs any ill will toward Walker for doing the same thing. One explanation for this is because they see him as white, not one of their own. American Indian scholar Duane Champagne notes that "light-skinned, blue-eyed individuals [claiming to be Indians] are viewed as white" by other Indians.[24]

Walker, like Nakia, often dresses entirely in blue denim as does Uncle Ray. This signifies his working-class roots—denim being the fabric of the hard-working cowboy, a figure who harks back to simpler times when "frontier justice" was swift. Like Nakia, Walker rides a horse for recreation, further embellishing his image as a cowboy. His horse's multicolored ("paint" or "pinto") coat, brown with large white patches, is a common film trope associated with American Indians. Uncle Ray also rides a paint horse and dresses entirely in denim. Like Walker, he too is both cowboy *and* Indian. The term "cowboy" also suggests a lone operator, a maverick, someone who refuses to be a team player (the same is true of the term "lone ranger"). What is more, as noted earlier, Walker's Cherokee name, Washoe, means "lone eagle." This is crucial because, like many other vigilante-type figures, Walker acts unilaterally, bending the law when he feels he needs to. However, unlike the Lone Ranger, who is an illegal vigilante, Walker, as an officer of the law, must work within the system. Like the Ranger, Walker is often portrayed in visual clichés that suggest either his own quasi-divinity as an avenging angel or that he is a messenger of God. These include backlighting, halo effect, being shot from low angles, framed against the sky, and the like.

WALKER AS "INDIAN-AT-WILL"

As Edward Said points out, the colonialist trope of a white man "passing" as native was a prominent theme in Lawrence's *Seven Pillars of Wisdom* (1922) as well as in Kipling's *Kim* (1901).[25] In colonial encounters it is easy for talented white men to pass as—even be accepted by—natives, yet it is difficult if not impossible for a native to be accepted as white. Indeed, as Cedric Clark argued in 1969, this theme of the assimilated native having to constantly prove his loyalty by becoming an enforcer of the dominant group's norms is nearly universal in the literature and media of colonialist countries, including the United States.[26]

An episode titled "Evil in the Night" (season 3; air date November 4, 1995), illustrates how Walker can become an Indian—a very powerful one—whenever it suits him to do so. An urban construction project unearths a Comanche grave site (the remains of women and children from the battle at Bandera Pass in 1841). Two white teenagers sneak into the excavation and disturb the remains. A shape-shifting Comanche named Running Wolf appears and summons the dead, and the boys are literally scared to death. Running Wolf has also played a role in the deaths of some officials involved in the construction project including a city councilman. The project is put on hold indefinitely, and officials feel the city is losing money. Once again, Indians are obstacles who impede "progress" by their very presence—even though these Indians are already *dead*. Walker seeks counsel from White Eagle, a Cherokee medicine man, who explains that Running Wolf is a "skin walker" who must avenge the death of his Comanche ancestors. Running

Wolf is depicted as a sort of Charles Manson–like character who enjoys killing. He has a snarky attitude and loose-flowing, shoulder-length hair (long hair in this series is often a signifier of criminality). Close-ups also reveal that Running Wolf has blue eyes. White Eagle cautions Walker not to tangle with Running Wolf because Walker has "lived too long among the whites. You have forgotten the tribal ways." Walker overconfidently responds that his father's Cherokee blood still runs through his veins. White Eagle agrees to tutor Walker. This is reminiscent of *Broken Arrow* (1950 and 1956), in which white savior Tom Jeffords takes a crash course on Apache culture, quickly mastering fighting Apache style. As Churchill notes, the white hero is "better at being an Indian than they [the Indians] are."[27] In preparation for his battle, Walker is seen taking part in a sweat-lodge ceremony inside a tipi and walking on hot coals (none of which are Cherokee rituals; scenes from "On Sacred Ground," March 11, 1995, are recycled freely here; see later discussion). Elders wearing Plains Indian headdresses participate, performing a sort of christening ceremony in which they give Walker a silver amulet. Walker finally confronts Running Wolf and kills him; this is notable, for Walker rarely if ever kills an adversary. This is very similar to an episode of *Law of the Plainsman* in which Apache deputy marshal Sam Buckhart must kill a "renegade" Apache named Gray Wolf, a childhood friend of Buckhart's.[28] Sometimes it becomes necessary for these characters to kill "one of their own" in order to demonstrate their depth of dedication to the dominant order.

WALKER AS "WHITE SHAMAN"

Cop shows have long been standard fare on U.S. television, but by the late 1960s, typical cop shows such as *Dragnet* had begun to seem stiff and perhaps even ridiculous to youthful tastes. If cop shows were to survive, they needed more interesting, "relevant" characters. Thus police dramas began incorporating "outsiders": women and minorities, even "hipster" white youths, as in *The Mod Squad* (ABC-TV, 1968–1973). Cordell Walker's Indian identity may also have been a way of deepening the psychology of the character, just as it did for Nakia Parker when, for example, Parker sits down to drink peyote tea with his "grandfather." Inner turmoil added psychological complexity to the often formulaic police/crime genre; at the same time it "positioned" these as distinct from other crime series: Walker is no ordinary cop. His Indianness helped make him unique, as it did for John Hawk and Nakia Parker (see chapters 6 and 7, respectively).

Walker incorporates ostensibly Indian religious rituals in order to deepen the psychology of the character as well as to make him appear more convincing as an Indian and, perhaps more importantly, to supply him with supernatural powers. The incorporation of these religious rituals into the series would create difficulties for Norris, however: for one thing, they offended Cherokees because they were entirely fake; for another, they offended some of Norris's "born-again" constituency who harbored deep hostility to "New Age" religion.[29] *Walker*'s writers freely invented religious rituals that have no analogue in native culture; some Cherokee scholars found this insulting, especially in view of the fact that Norris himself claims to be half Cherokee.[30] For example, in the second episode ("Borderline") Walker performs a spurious Cherokee ritual at his fallen partner's grave.

Walker dons a red serape that looks very much like Superman's cape (figure 8.3). The camera angle could scarcely get any lower; he is framed against the sky in a godlike pose, with flute music in background. Had Norris bothered to research his purported Cherokee heritage—or hired a Cherokee consultant for the series—he might have discovered that an observant Cherokee would not perform a native ritual on a non-Cherokee's grave. Cherokee scholar Richard Allen, the Vine Deloria Jr. Distinguished Indigenous Scholar and policy analyst for the Cherokee Nation, explains, "[T]raditional Cherokees believe the owl to be a harbinger of death or ill-omen and would find the

Figure 8.3. Walker transforms himself into a shaman by donning a red serape that is equivalent to Superman's cape.

owl's feather repulsive."[31] Robert Conley, professor of Cherokee studies at Western Carolina University, adds, "The owl is usually thought to be a *sgili* (a witch) in disguise."[32] Suddenly, the camera shifts to a very high angle, a "God's eye view." The viewer is now positioned as God, looking down on his messenger. God hears his prayer:

> **Walker:** (Holding up a feather, looking like a priest holding a crucifix) My friend, go with the spirit of the owl. He will see all things for you.
>
> **Walker:** (Holding up a second feather) Go with the spirit of the hawk. He will protect you on your final journey. (Holding up a third feather) Go with the spirit of the eagle. He will soar you [*sic*] to your father. Your father awaits you.

The coloring of the feather suggests Walker's hybrid character: he too is mostly white with a bit of brown; his Indianness emerges when hypermasculinity is required. Walker gazes up at the feather intently. Crouching down, Walker silently places all three feathers on the grave. His black hat is in a direct line with the other gravestones, almost as if it were one, too.

> **Walker:** Payback time.

Now it is time for some old-fashioned "frontier justice." "Frontier justice" means no arrests, no reading of rights, no courts—the vigilante becomes judge, jury, and executioner. Of course, being a legitimate lawman, Walker can do no such thing, but he may go berserk and beat the perpetrator within an inch of his life.[33] Thus vengeance, a popular theme in westerns, is dignified through ritual. What transpires in this ritual is more or less what occurs when Clark Kent steps into a phone booth: stripping off his shirt and tie—symbolizing his Anglo-American identity—the mild-mannered Kent transforms himself into a hypermasculine avenger.[34] Walker's serape serves the same function as Superman's cape, which is also red. Thus, like Cochise or Sam Buckhart, Walker becomes a superhuman Indian enforcer of Anglo-American norms (see chapters 3 and 4). Calling on his native half, he allows himself to go berserk, like the Hulk or Rambo. His white half is no longer responsible for his actions; all culpability for his overzealous (and extralegal) behavior is displaced onto his Indian alter ego.

WHITE SHAMANISM: BACKGROUND AND HISTORY

"White shamanism," a term coined by Geary Hobson, is the practice of whites adopting and synthesizing native and Eastern spiritual principles (the term "shaman" comes from Siberia). According to Hobson, white shamanism began in 1970s U.S. literature with a group of poets who felt they had achieved a sort of native spirituality through the use of peyote. These practices later agglomerated into a hodgepodge of Eastern and American Indian philosophies that came to be called New Age spiritualism.[35] This sounds innocent enough at first and may even seem like an homage to American Indians; however, it has more insidious implications for Indians themselves.

Walker can become a native anytime he likes by performing a quasi-native ritual or simply by putting himself in a trance. In an episode titled "On Sacred Ground" (season 2; air date March 11, 1995), Walker becomes a shaman. Walker, deep in the forest, dons an Apache-style bandanna like the one worn by Sam Buckhart in *Law of the Plainsman* (figure 8.4). His face is lit mostly on one side by the fire he contemplates. His head is backlit by moonlight, creating a halo effect. A montage takes shape: in a series of dissolves, three images fade into one another: Walker's face, then the fire, and then flashbacks of some unspecified American Indians (probably Plains Indians) doing a sacred dance to quasi-traditional music. Colors become increasingly hotter as the camera (i.e., the viewer) draws closer to the fire. The viewer now sees inside Walker's head, witnessing visions of his ancestors. This transformative ritual again is analogous to Clark Kent's stepping into a telephone booth. Walker continues to meditate by the fire, putting himself in a trancelike state, like a yogi. His eye line does not match the previous shots: when the camera violates the 180-degree rule, it suggests a godlike point of view, since a human being cannot be in two places at once.

The backlighting (simulated rays of moonlight) indicates Walker's now-supernatural state (figure 8.4). It should be also noted that the display of Walker's torso correlates with his native phase; he is rarely if ever seen bare chested otherwise (see discussion of male bodily display in chapter 3 on *Broken Arrow*). The unsaddled horse in the background on the right confirms Walker's Indian state—"Hollywood Indians" generally ride bareback. After achieving the proper mental state, Walker proceeds to walk on hot coals. This is not and never has been an American Indian practice.[36] The program's producers here have conflated American Indian culture with Hindu culture in a sort of New Age stew: one could say they literally got their Indians mixed up. The issue here is not lack of verisimilitude or that television writers do not conduct proper research but that they simply make Native Americans into anything they want them to be. In the 1990s, "fire-walking," promoted by Robert Bly in his book *Iron John*, was a ritual used to bolster men's sense of masculine prowess.[37] *New York Times* reporter Richard Schweder asserts that this was a reaction to a "crisis of masculinity" triggered by the threat of feminism.[38] A contemporaneous article in *Newsweek* noted that white males felt their formerly secure identities threatened by "feminists, multiculturalists, affirmative-action employers, rap artists, Native Americans."[39]

Figure 8.4. Rays of moonlight illuminate Walker from behind as Walker prepares to "fire-walk." Walker is only seen shirtless when he is in Indian mode.

As stated earlier, white shamanism might at first seem like a tribute to American Indians, but the implications go much deeper than Hollywood's "made-up mumbo jumbo" and cultural ignorance regarding native spiritual practices. This is a serious issue among native scholars, many of whom view it as cultural theft and worse. Jimmy Durham writes,

> There is in the United States . . . a curious phenomenon that is seldom given intellectual consideration: whites claiming to be part Indian and, even more, whites who claim to be Indian. Surely there is not another part of the world wherein members of the racist oppressor society claim to be members of the oppressed group.[40]

Apparently, it is not enough for Euro-Americans to colonize American Indians' land and resources—they must also appropriate their culture, thereby naturalizing their claim to the land and its bounties. In the words of Margo Thunderbird,

> They came for our land, for what grew or could be grown on it, for the resources in it, and for our clean air and pure water. They stole these things from us. . . . Now they want our pride, our history, our spiritual traditions. They want to rewrite and remake these things, to claim them for themselves.[41]

Some in fact view this as the final stage in the process of complete colonization. John Lavelle, director of the Center for Support and Protection of Indian Religious and Indigenous Traditions, adds, "This is the final phase of genocide. . . . First whites took the land and all that was physical. Now they're going after what is intangible. . . . It's the conqueror fantasizing about who he has conquered."[42] By pretending to be natives, the colonizers might be able to convince themselves they are the rightful and true possessors of the land and its heritage. Apparently Anglo-American men subconsciously (or even perhaps consciously) realize they lack authenticity as rulers of the land. Daniel Francis explains,

> [Euro]-American civilization has . . . always been uncomfortable with our treatment of the native peoples. More than that, we have also suspected that we could never be at home in America because we were not Indians. . . . One way non-natives choose to resolve this anxiety is to somehow become Indians.[43]

What is more, they may feel insecure in their masculinity, so they appropriate the masculine prowess and superhuman powers they have traditionally attributed to American Indians. Thus they (mis)appropriate attributes of natives in forging their *own* myth; then they reiterate this myth to each other in films and television, where it takes on an aura of quasi-authenticity, or what Aumont calls "plausibility."[44] *Walker*'s America, like Disneyland, is a *simulacrum* in which white males not only reign supreme but are better at being Indians than Indians themselves. "These are non-Indians who think they can be better Indians than we are," said Carol Standing Elk of the American Indian Movement.[45] White men can become Indians easily enough, yet it is extremely difficult if not impossible for Indians such as Sam Buckhart or Nakia Parker to become white. Walker overrides this difficulty by being both white *and* Indian.

WALKER/NORRIS INTERFACE AND ISSUES OF INDIANNESS

There is a good deal of spillage between character Cordell Walker and the public persona of Chuck Norris; in fact, it is sometimes unclear where one ends and the other begins. As mentioned earlier, Norris brought his own experiences and views to the series. In the first instance, both Norris and Walker take pains to assert they are half Cherokee. In both his autobiographies (1998 and 2006), Norris makes an issue of being a half-breed, as does the character he plays in *Walker*. In season 3, episode 6 ("Rainbow Warrior") Walker calls himself "a member of this tribe," the Oklahoma Cherokees.[46] In the made-for-TV movie that served as the pilot for the series, Walker, in flashback, is seen as a tow-headed child being bullied by a gang of native children. This is taken from Norris's childhood experience of attending school with Indian children in Arizona who bullied him.[47]

Reading Norris's autobiographies (1988 and 2006), one cannot help noticing his ambivalence toward American Indians. Despite his own and his character's repeated claims to Indianness, Norris works hard to distance himself from his Cherokee roots—and from his father. Norris was appalled by his father's erratic behavior and tended to disassociate himself not only from his father but even from his father's side of the family. According to Norris, his father had dark skin, "coal-black" eyes, and black hair. In other words, he looked like an Indian. Norris writes that his "most difficult and confusing relationship" was with his father and refers to his father "a negative role model." Complicating matters, Norris's father was an abusive alcoholic who apparently fit the drunken-Indian stereotype. It is evident that the role of Uncle Ray is partly based on Norris's father, whose name was Ray (Norris's middle name is Ray as well). Uncle Ray, a full-blood Cherokee, is often made to appear silly and ineffectual; like Tonto in *The Lone Ranger*, he is emasculated, feminized. What is more, he is trivialized. Ray sometimes hangs out in the neighborhood bar acting tipsy and telling inane jokes.

Ray is frequently the butt of ridicule. For example, while Walker is viewing a videotape of a bank robbery from a closed-circuit camera, Ray asks, "What's on? I missed *People's Court* last night" (figure 8.5). This comment makes Ray sound silly and uneducated, having extremely pedestrian tastes in television (perhaps a bit of postmodern irony here). Ray is in the background, out of focus. He is nearly invisible, both literally and figuratively.

Like Norris himself, Walker tries to distance himself from his Cherokee heritage, but Ray will not let him. Ray (figure 8.6) is framed against a tree and the sky, symbolizing his connection to nature, while Walker stands in front of his house, symbolizing civilization and the comforts of modernity.

> **Uncle Ray:** Washoe, you forgot to say good morning to the sun.
>
> **Walker:** (looking up, reluctantly) Morning . . .

Walker egregiously misrepresents Cherokee culture. Cherokee scholars have expressed displeasure with *Walker* and hold Norris responsible. Richard Allen asserts, "Norris distorts not only Cherokee but the broad sweep of Indian culture."[48] Robert Conley calls *Walker*'s portrayals of Cherokee culture "absurd and insulting," adding, "Norris seems to know nothing about Cherokee

Figure 8.5. Uncle Ray, wearing denim, is marginalized, in the background. Note Navajo patterns on Walker's shirt.

culture. Furthermore, he does not seem to give a damn enough to try to find out."[49]

Ultimately, Cordell Walker deploys his Indianness when it helps him solve crimes but otherwise tries to distance himself from it. This could be said of Norris himself, who employed his (ostensible) Indianness to advance his career but generally distances himself from it. In fact, Norris's professed hero is John Wayne, who portrayed Hollywood's most vehement (and violent) Indian hater in John Ford's *The Searchers.*[50] In what could be deemed a "Freudian slip," Norris, who now styles himself a political commentator, explains what the United States' founders had in mind when "*we* came over from England."[51] His use of the pronoun "we" with reference to the English settlers clearly negates his claims of Indianness, indicating that he sees himself as thoroughly Anglo-American.

Figure 8.6. Ray is associated with nature.

FUNDAMENTALIST RELIGIOUS OVERTONES AND GENDER ROLES

Norris himself is a born-again Southern Baptist and self-described "conservative Christian."[52] This is worth examining because of the conflicts that arose between *Walker*'s New Age mumbo jumbo and conservative Christianity. Although many Cherokees have been raised as Southern Baptists and have managed to reconcile their traditional beliefs with Christianity, this transfer does not work both ways.[53] Some conservative Christians took offense to *Walker*'s New Age spiritual practices.[54] By season 4 these were abandoned in favor of references to Christianity because Norris himself was uncomfortable with them.[55] At this point Norris had "recommitted" himself to evangelical Christianity.[56]

Religion is germane to this discussion because fundamentalist Protestants tend to embrace strict gender roles and espouse rigid ideas about masculinity and what it means. For example, many conservative Christians subscribe to the biblical view that women should accept subservient roles.[57] Some of these strictures are embedded in *Walker*: for one, there are no women police officers until season 7. Conservative Christians also tend to oppose civil rights for gays.[58] This is directly related to issues of American masculinity and what these mean. What is more, many conservative Christians label themselves "pro-life" while celebrating militarism and war—this is especially true of Norris himself, whose earlier careers playing the militaristic Colonel Braddock and Major McCoy demonstrate. Both Braddock and McCoy prefer violence—sometimes on a mass scale—to negotiation.[59]

Fundamentalist Protestant ideals have always been deeply imbricated in the western genre. That fundamentalist longings are at the heart of the western has been pointed out by such scholars as J. Fred MacDonald, Peter Homans, and Michael T. Marsden.[60] Marsden writes that the western—in all its formats, whether in novels, film, or television—is at base a variation of a Puritan morality play. However, the western tale's religious overtones are no longer overt, although these may be apparent to fundamentalists who recognize them. Marsden explains,

> For the Puritans, the wilderness of their "Chosen Land" was inhabited by devils [Pequots], and these devils could be driven out only by the strongest and worthiest of men. . . . It is of this challenge that the American hero, beginning with Daniel Boone, was born. And Boone's cultural descendant, the western hero, became America's permanent heroic creation, serving as Redeemer for generations of Americans.[61]

Finally, conservative Christians take a biblical view of the land and how it should be used, which is antithetical to the way most Indians feel about the land. Ann Coulter, a conservative author and fundamentalist Protestant, explained this *dominionist* view in 2001 on Sean Hannity's Fox News program:

> COULTER I take the biblical idea. . . . God gave us the earth. . . . God says, "Earth is yours. Take it. Rape it."
>
> PETER FENN Terrific. We're Americans, so we should consume as much of the earth's resources—as fast as we possibly can.
>
> COULTER As opposed to living like the Indians.[62]

Norris identifies with conservative-Christian views, and by season 4 he began actively incorporating these into *Walker, Texas Ranger*. Norris himself envisioned the series as a Christian program: he wrote in his 2006 autobiography that the series indeed contained an evangelical tone and went on to boast that it was voted Best Christian Program at the 1998 Epiphany awards.[63]

POLITICAL IMPLICATIONS

Along with subtle and not-so-subtle religious themes, Norris brought his conservative political stance to the series. *Walker* had clear political implications: like the Lone Ranger, this "fabulous individual" set the tone for how U.S. leaders should behave: "cowboy justice" was the ideal. *Walker* may have been part of a "swing to the right" that transpired in the late 1990s and early 2000s. *Washington Post* columnist Paula Span writes that *Walker, Texas Ranger* "helped land George W. Bush in the White House" in 2000 (Bush ran for governor of Texas one year after *Walker*'s debut). "Is it so far-fetched that last November a significant proportion of [voters] looked at George W. and thought, 'Hey, we know this guy'?"[64] Walker and Bush have several characteristics in common: both are from Texas, both affect a cowboy image, both style themselves as "compassionate conservatives" and conspicuous Christians, both portray themselves as tough but fair.[65] This image, exemplified by Cordell Walker, a tough, no-nonsense western hero, is so embedded in U.S. culture that many voters think of this character as a prototype and that their country's president should behave like him, a hero "who cites scripture about forgiveness while unremorsefully thwacking away at enemies foreign and domestic."[66]

THE INDIANIZED WHITE MAN AND ANGLO-AMERICAN MASCULINITY

Like its protagonist, the series itself is a hybrid, both a crime series and a western. Tropes of Anglo-American masculinity and patriarchy thrive not only in westerns but in police dramas. As in most westerns, issues of masculinity—and challenges to it—abound. The function of these texts, at base, is to reassure the public that ruthless men stand prepared to protect us from even more ruthless men, that it takes a hypermasculine male (or an avenging angel) to protect us from other hypermasculine males.

The earliest images of the New Man liberally borrowed traits ascribed to Indians. Like the Indian, the New (Euro-American) Man is a man of the wilderness. U.S. literature and media are riven with such men of the wilderness, "mountain men," and the like. At some point, probably when the western began to wane in popularity in the 1970s, the wilderness migrated to the "urban jungle," wherein "fighting crime" was posed as a war with barbarian Others who lived here in our cities.[67] This urban-jungle theme and its concomitant ruthless police-protector protagonist became manifest in *Dirty Harry* (1971) and on television in *Kojak* and was perhaps spelled out most clearly in the 1981 film *Fort Apache, the Bronx.* Instead of simply relocating to the new, "urban frontier," *Walker*—as did *Nakia* and a couple other David

Gerber TV series, such as *Cade's County*—overtly combined elements of the western with the urban-jungle theme.

The New Man's strength is visually derived from his connection with the wilderness.[68] This is apparent not only in TV programs but also in advertisements. Walker's resemblance to the Marlboro Man (figure 8.8) can hardly be coincidence (note denim jacket symbolizing the workingman). The Marlboro Man's head is carefully framed against the mountains, lending the strength of the wilderness to the character, and his hat is connected to the sky. In the

Figure 8.7. Cordell Walker stands guard outside Dallas.

Figure 8.8. The Marlboro Man from a 1974 magazine ad. Cigarettes were coded as masculine.

metropolitan variation (figure 8.7), the cityscape substitutes for mountains. However, Walker is standing *outside* the city: though he works to clean up crime in the city, he is not *of* the city. His shirt, with its Navajo designs, connotes his Indianness *beneath* his cowboy garb. Walker is shot from an extremely low angle, framed against the sky, making him appear godlike.

CONCLUSION

Cordell Walker embodies a highly evolved fusion of European and American Indian masculinities, refining a process that began in U.S. literature in the 1820s if not earlier. His Indianness is nonobtrusive, covert: he can don his Indian secret identity—like a superhero's cape—any time he needs superhuman powers (i.e., hypermasculinity). Like Cherokee chief Mingo on *Daniel Boone*, Walker is *both* white and Indian; he is the best of either world at any given moment, as it suits him. His white half makes him serious, patient, linear thinking, analytical, but he is not "locked into" Enlightenment rationality or logic.[69] His Indian half makes him stronger, more spiritual, more natural, as well as—like John Rambo, who was later revealed to be half Indian—impetuous, apt to go berserk when and if the need arose.[70] Walker, then, is the perfect Good Indian: a temporary one. He does not overstay his welcome, nor make demands, nor remind Anglo-Americans that he is the native and they are not. When Walker's inner Indian is no longer useful, it goes back into hibernation until, like a genie, it is summoned again. Thus "the only good Indian" is one who appears on command—to help enforce Anglo-American law and order—and recedes when not needed.

NOTES

1. Vine Deloria Jr. in Gretchen M. Bataille and Charles L. P. Silet (eds.), *The Pretend Indians: Images of Native Americans in the Movies* (Ames: Iowa State University Press, 1980), 189.
2. Drinnon, 1980, 46–57.
3. Edward Said, *Culture and Imperialism* (New York: Vintage Books, 1993), 106.
4. Philip J. Deloria, *Playing Indian* (New Haven, CT: Yale University Press, 1998), 2.
5. Chuck Norris with Ken Abraham, *Against All Odds: My Story* (Nashville, TN: Broadman & Holman, 2006), 35–36.
6. Rescuing missing prisoners of war in North Vietnam had been the premise of the film *Rambo: First Blood Part II* (1985).
7. Mike Hughes, "Chuck Norris Finds Television and Fatherhood Better Second Time Around." *Tulsa World*, October 16, 2005, TV5.
8. Cordell Walker appeared in two other series: *Martial Law* (CBS, 1998–2000), a short-lived police drama in which the protagonist practices martial arts, and *Sons of Thunder* (1999), a spin-off of *Walker*, produced by Norris and his brother Aaron. Walker returned in a 2005 made-for-television movie, *Walker, Texas Ranger: Trial by Fire*. The original *Walker, Texas Ranger* still appears in reruns on the USA Network and the Hallmark Channel. Norris is also featured in a video game called *Chuck Norris Superkicks*.
9. Ann Hodges, "Money Woes May Sideline Norris Series." *Houston Chronicle*, April 14, 1993, 1.
10. Norris, 2006, 8. Norris, who had blond hair as a youth, claims his maternal grandfather was full-blooded Cherokee (15). However, according to the page titled "Carlos Ray Norris

a.k.a. 'Chuck Norris' Family History" on the RootsWeb website, his maternal grandfather, John Porter Scarberry, had blue eyes (http://wc.rootsweb.ancestry.com/cgi-bin/igm.cgi?op=GET&db=14x24&id=I0141&style=TABLE). The chances of a full-blood Cherokee having blue eyes are less than miniscule: "Blue eyes only occur in people who have blue-eyed Caucasian relatives on *both* sides of their family tree, and even then only some of the time." Orrin Lewis, "Setting the Record Straight about Native Peoples." NativeLanguages.org. http://www.native-languages.org/iaq10.htm.

11. The motif of American Indian characters with military backgrounds is fairly common in films and television and includes Chief in *Garrison's Gorillas* (ABC, 1967), Billy Jack in *Billy Jack* (1970), Rambo in *First Blood Part II* (1985), and Walker himself, who is depicted in flashbacks as a former Army Special Forces operative (Green Beret) in Vietnam. American Indians have indeed served in inordinately high proportions in the U.S. military, especially in commando units. Tom Holm, "Patriots and Pawns: State Use of American Indians in the Military and the Process of Nativization of the United States." In M. Annette Jaimes, ed. *The State of Native America: Genocide, Colonization, and Resistance* (Boston: South End Press, 1992), 345–70.

12. Even as a token character, Uncle Ray does not last long: he dies after fourteen episodes.

13. George Gerbner, Casting and Fate: Women and Minorities on Television Drama, Game Shows, and News. Screen Actors Guild report, 1993. http://www.asc.upenn.edu/gerbner/Asset.aspx?assetID=1074.

14. Other American Indian characters associated with martial arts are Billy Jack (*Billy Jack*, 1970) and Rambo (*First Blood Part II*, 1985).

15. This is yet another stereotype: "Characters with mental disabilities were 4–1/2 times more likely to be criminals or villains and three times more likely to commit violence." Gerbner, 1993, 9.

16. Will Wright, *Six Guns and Society* (Berkeley: University of California Press, 1977), 52, 80.

17. Ward Churchill remarks that "the only thing observably 'Indian' about Walker is that he stops in every episode to visit with his Uncle Ray." Churchill, *Fantasies of the Master Race: Literature, Cinema, and the Colonization of American Indians* (San Francisco: City Lights Books, 1998), 221, n. 179. This, however, is not entirely the case.

18. *Walker, Texas Ranger: The Complete First Season* (Paramount/CBS Home Video, 2006). DVD.

19. Bob Kane, "Legend: The Bat Man and How He Came to Be" *Detective Comics* 33, 1939.

20. Matthew A. Stern, "Review: The Best of *Bravestarr*." PopMatters. July 18, 2007. http://www.popmatters.com/pm/review/the-best-of-bravestar.

21. Ward Churchill, *Struggle for the Land: Native North American Resistance to Genocide, Ecocide, and Colonization* (San Francisco: City Lights Books, 2002), 274–79.

22. "Museum Archives: FAQ's." Museum of the Cherokee Indian. http://www.cherokeemuseum.org/html/archives_FAQb.html. As Richard Conley suggests elsewhere, one might reasonably assume Norris would know enough about Cherokee culture to avoid such errors.

23. There is a Cherokee reservation near Dallas whose land grant was recognized by Spain in 1807. See "About Sovereign Cherokee Nation Tejas." Texas Cherokee Nation. 2010. http://www.texascherokeenation.org/about.php.

24. Duane Champagne, *Contemporary Native American Cultural Issues* (Lanham, MD: Rowman & Littlefield, 1999).

25. Said, 1993, 161.

26. Cedric C. Clark (Syed M. Khatib), "Television and Social Control: Some Observations on the Portrayals of Ethnic Minorities." *Television Quarterly*, Spring 1969, 18–22.

27. In *Images of Indians: How Hollywood Stereotyped the Native American*, directed by Chris O'Brien and Jason Witmer (Starz Encore Entertainment, 2003).

28. *Law of the Plainsman*, episode 6: "Blood Trails" (Cardiff Productions/Four Star Television/NBC-TV. Air date: November 5, 1959).

29. Sheila Overturf, "Program Reviews: *Walker, Texas Ranger*." ChristianAnswers.net, 2002. http://www.christiananswers.net/spotlight/tv/2002/walkertexasranger.html.

30. Richard Allen, PhD, e-mail correspondence with author, May 6, 2010; Robert Conley, e-mail correspondence with author, May 6, 2010; also see Mercurie, "The Invisible Minority: Native Americans on American Television." *A Shroud of Thoughts* (blog). July 6, 2009. http://mercurie.blogspot.com/2009/07/invisible-minority-native-americans-on.html.

31. Allen, e-mail, 2010.

32. Conley, e-mail, 2010.

33. Kirby Farrell, "The Berserk Style in American Culture." *Cultural Critique* 46, Autumn 2000, 179–209.

34. Jerry Siegel and Joe Shuster, *Action Comics* 1, June 1938. Excerpted at "Religion of Superman/Clark Kent." Comic Book Religion. http://www.comicbookreligion.com/?c=1& Superman_Clark_Kent; *The Adventures of Superman* (Motion Pictures for Television, 1952–1958).

35. Geary Hobson, *The Remembered Earth: An Anthology of Native American Literature* (Albuquerque: University of New Mexico Press, 1981), 104.

36. Allen, e-mail, 2010; Conley, e-mail, 2010.

37. Robert Bly, *Iron John: A Book about Men* (Boston: Addison-Wesley, 1990).

38. Richard A. Schweder, "What Do Men Want? A Reading List for the Male Identity Crisis." *New York Times*, January 9, 1994. http://www.nytimes.com/1994/01/09/books/what-do-men-want-a-reading-list-for-the-male-identity-crisis.html. *Firewalker* was also the title of a 1996 Norris action film released by Cannon.

39. David Gates, "White Male Paranoia." *Newsweek*, March 29, 1993, 48–53.

40. Jimmy Durham, "Cowboys and . . ." In M. Annette Jaimes, ed. *The State of Native America: Genocide, Colonization, and Resistance* (Boston: South End Press, 1992), 424.

41. Quoted in Wendy Rose, "The Great Pretenders: Further Reflections on White Shamanism." In Jaimes, 1992, 403.

42. However, it should be noted that there are some Indian religious practitioners who feel that the native ways should be shared with and by anyone who appreciates them. Dirk Johnson, "Spiritual Seekers Borrow Indians' Ways." *New York Times*, December 27, 1993, A1.

43. Daniel Francis, "The Imaginary Indian." In Tania Das Gupta, Carl E. James, Roger C. A. Maaka, Grace Edward Galabuzi, and Chris Anderson, eds. *Race and Racialization: Essential Readings* (Toronto: Canadian Scholars Press, 2007), 239.

44. Aumont et al., 1992, 117–18.

45. Johnson, 1993, A1.

46. However, Norris himself is *not* a registered member of the Cherokee Nation nor the United Keetoowah Band; if he were half Cherokee, he would be eligible to join either. Allen e-mail, 2010.

47. Norris, 2006, 15–16.

48. Allen, e-mail, 2010.

49. Conley, e-mail, 2010.

50. Chuck Norris with Joe Hyams, *The Secret of Inner Strength: My Story* (Boston: Little, Brown, 1988), 206. Norris's idol John Wayne was not an admirer of Indians in real life, either. In 1971 he told interviewer Richard Warren Lewis, "I believe in white supremacy." Asked specifically how he felt about Indians and their loss of the land, he responded, "I don't feel we [Euro-Americans] did wrong in taking this great country away from them. . . . Our so-called stealing of this country from them was just a matter of survival. There were great numbers of people who needed new land, and the Indians were selfishly trying to keep it for themselves." Richard Warren Lewis, "Candid Conversation: John Wayne." *Playboy*, May 1971, 78. Reprinted in Barry G. Golson, ed. *The Playboy Interview* (New York: Playboy Press, 1981), 203–27.

51. B&H Publishing Group, "Black Belt Patriotism: Chuck Norris." YouTube. http://www.youtube.com/watch?v=kMhLgC8WGz8; emphasis added.

52. Norris, 2006, 22.

53. Allen, e-mail, 2010.

54. Overturf, 2002.

55. Norris, 2006, 181–82.

56. Chuck Norris, "Saving a Roundhouse for Romney." *World Net Daily*, January 7, 2008.

57. As delineated in Ephesians 5:22 and 6:5–6; Colossians 3:18; Titus 2:3–5; and Peter 3:1. New American Standard Bible (Grand Rapids, MI: Zondervan, 2011).

58. See Leviticus 18:22, New American Standard Bible, 2011.

59. James Kendrick, *Fighting Outward, Looking Inward* (Carbondale: Southern Illinois University Press, 2009), 112–13.

60. Fred J. MacDonald, *Television and the Red Menace: The Video Road to Vietnam* (Westport, CT: Praeger, 1985), 139–40; Peter Homans, "Puritanism Revisited: An Analysis of the Contemporary Screen-Image Western." In Jack Nachbar, ed., *Focus on the Western* (Englewood Cliffs, NJ: Prentice-Hall, 1974), 84–92; Michael T. Marsden, "Savior in the Saddle: The Sagebrush Testament." In Nachbar, 1974, 93–100.

61. Marsden in Nachbar, 1974, 93.

62. *Hannity & Colmes* (Fox News Network, June 20, 2001). Following Coulter, Norris has become a political and cultural commentator. He has his own column in the conservative *World Net Daily*.

63. The awards are sponsored by the John Templeton Foundation. Norris, 2006, 226.

64. Paula Span, "That Lone Star Good Guy: George Bush? Nope. Walker, Texas Ranger." *Washington Post*, May 19, 2001, C1. Mark Crispin Miller also mentions in 1986 that "television's contemporary dispenser of justice shares a niche with the president of the United States. "Deride and Conquer." In Todd Gitlin, ed. *Watching Television: A Pantheon Guide to Popular Culture* (New York: Pantheon Books, 1986), 155.

65. Coincidentally, George W. Bush's middle name is Walker.

66. Span, 2001, C1.

67. In "Sitcoms and Suburbia: The Role of Network Television in the De-urbanization of the U.S." (unpublished master's thesis, University of Florida, 2007), I argue that this distorted image of the city was part of the reason for the mass exodus to the suburbs in the U.S. during this period.

68. Wright, 1977, 52, 80.

69. Philip Deloria, 1998, 3.

70. See Farrell, 2000, 179–209. Rambo's Indianness was not revealed until the sequel, *Rambo: First Blood Part II*. It was not part of author David Morrell's original conception of the character. David Morrell, e-mail correspondence with author, January 19, 2009.

Chapter Nine

Televisual Quasi-history

Retrospective Justification for Conquest

> *Pleasure can be manipulated by or at least articulated to repressive forms of power and existing structures of inequality. . . . Pleasures may themselves be repressive and regressive—for example those derived from relations of domination over other groups in formations of racism.*—Larry Grossberg.[1]

This chapter will delve into some of the underlying psychological mechanisms and strategies at work, including the complicity of television viewers themselves, an issue that should not be understated. It will review the visual strategies examined in this study and articulate how these operate within a historical system of race-based representation that has survived from the original Pocahontas myth all the way through novels, films, and finally television. These tropes have been remarkably consistent throughout U.S. history but at the same time malleable enough to apply in various periods in various ways. Most of the historical analyses in this study have focused on the Cold War era because that is the period that coincides with the nationwide expansion of network television, but these themes have been remarkably consistent for centuries.

The most crucial yet most elusive point is that televisual representations of American Indians have rarely been recognized as a continuing form of *colonialist discourse*. Colonial-discourse theory explains the motivations and benefits of racist and xenophobic representations: one group is intent on colonizing the land and resources of the other but requires an elaborate rationale in order not to have to face that its members are conducting themselves in a manner that contradicts their own ethical or religious beliefs, as Montesquieu pointed out in 1748.[2] It is not as simple a proposition as "us versus them," as the Plymouth Colony Pilgrims tended to construct this dichotomy. Although in retrospect it seems obvious that the Pilgrims *needed* to despise the Pequots in order to justify their own cruelty, they nonetheless admired the natives in many ways and appropriated some of their ideas and philosophies.

In the early days of the republic (and even before that), a third position, a synthesis of Us and Other materialized: this was the Indianized white man, who combines the best qualities of both races. Indian scholar Philip J. Deloria notes, "Through Indianness, colonists articulated a revolutionary identity. . . . Reflecting the colonists' need to define themselves as something new and non-British."[3] The historical Daniel Boone and the fictional Natty Bumppo were early examples of this figure, but it has thrived throughout centuries—modified, of course, to suit various objectives and periods.[4] This model continues to address uniquely American concerns, whatever these might consist of at any given point. Indeed, this figure is so versatile it has been used both to glorify U.S. society as well as to critique it. The Indianized white man has gone from novels to film to television, where it was exemplified by Cheyenne Bodie in *Cheyenne* (1955–1963), Tom Jeffords in *Broken Arrow* (1956–1960), Hondo Lane in *Hondo* (1967), and Cordell Walker in *Walker, Texas Ranger* (1993–2001). Sometimes this synthesis is approached from the opposite direction, as the Europeanized (Anglicized) Indian: Tonto in *The Lone Ranger* (1949–1957), Cochise in *Broken Arrow* (1956–1958), Sam Buckhart in *Law of the Plainsman* (1959–1960), John Hawk in *Hawk* (1966), and Nakia Parker in *Nakia* (1974). The Indianized white man and the Anglicized Indian are intimately related, almost—but not quite—identical. When the character is a literal hybrid (i.e., a half-breed), he generally identifies with the Euro-American side with only part-time identity as an American Indian.

Although Clark's four-stage model provides an organizing framework for the current study and has much explanatory power, in some respects it tends to be a bit simplistic and requires more flexibility and nuance. Even if Clark's assumptions are generally correct, which many appear to be, there are no overriding or foregone conclusions that can be taken for granted, and there are always exceptions. This is reflected in the fact that the visual operations in the programs studied herein are often contradictory; in addition, the visuals frequently undermine the dialogue, sometimes even the narrative itself. Stuart Hall asserts that identity operates on a contested terrain that media owners can and do dominate *but cannot fully control* due to the fact that they must make their products attractive to various segments of society in order for them to become popular—in every sense of the term—and therefore profitable.[5]

In chapter 1, a survey of American Indian representations revealed a few distinct motifs, most notably that of the Indian enforcer (Clark's "Regulator" stage). Another trend I noted was what Clark/Khatib calls "temporal displacement," which entails situating the character(s) in a different era in order to avoid confronting viewers with current controversies or making viewers uncomfortable lest they associate their discomfort (i.e., cognitive dissonance) with the product(s) being advertised. Chapter 2 examined Tonto in *The Lone Ranger* (1948–1957). Tonto is clearly based on one of the archetypal literary figures: the "Indian companion" who assists the Euro-American conqueror—or savior, depending on whose ideology is at stake—in his task of bringing order to the wilderness. It seems clear that this Indian-companion figure evolved from the legend of Pocahontas, which morphed into Defoe's Friday (1719) and Cooper's Chingachgook (1820s) and kept going, right on into the twentieth century. Tonto is a certainly a Regulator in that he helps the white patriarch enforce Euro-American norms. He is also feminized, almost like a wife to the Ranger: his resemblance to Pocahontas in this regard is clear. Tonto's innate ability to connect with the land complements the Ranger's position as an exalted moral figure or divine avenger. This Indian–white pairing combines the earthy instincts of the Indian with the intellect of the Euro-American, a perfect team. Tonto became a willing helper in the Ranger's quest to make the land—(mis)represented as an empty space waiting to be civilized—safe for Euro-American settlers. Even though Tonto is never directly denigrated, his overall depiction is nonetheless a racist stereotype: the racist position has merely shifted from one of blatant hatred, common to canonical westerns, to one of condescending paternalism. Historian Christopher Lasch notes, "Racism and paternalism were two sides of the same coin."[6] In either case the object being managed is deemed subhuman. This shift from the canonical Bloodthirsty Savage to Noble Savage would become a model for American Indian portrayals in the post–World War II revisionist paradigm. This worked to ameliorate the problem of racism in the Cold War arena by finding "a third way out" (Barthes's term).

Chapter 3 examined *Broken Arrow* (1956–1958), which takes the Regulator a step further, situating the Apache as an ostensible equal to the Anglo-American savior. We also see these two figures becoming more like each other, meeting in the middle, as it were. Cochise becomes Superfink, the Regulator with superhuman (or near-superhuman) qualities, who enables and assists in the Anglo-American control of his territory (*Brave Eagle* had done this in 1955, but he did not possess superhuman capabilities). The Anglo-American savior (Tom Jeffords) is slightly humbled and presented as humanitarian rather than as conqueror. As explained in the introduction, diplomatic necessities associated with the Cold War partly dictated this moderation. The Indian leader is characterized as a *junior* partner in the enterprise of Euro-American colonization. Chapter 3 also examined issues of masculinity and the ways in which the Other is fetishized, feminized, sexualized. For example, the character of Cochise is alternately—sometimes simultaneously—

depicted as hypermasculine *and* as feminine. Another aspect of the series I was able to uncover was how *Broken Arrow* served as a public-relations vehicle for the U.S. government, helping to elide a contemporaneous political controversy that had direct and dire implications for real-life American Indians: the unilateral termination ("abrogation" would be a more accurate term) by Congress of longstanding treaties with certain, selected sovereign Indian nations. I discovered written evidence that Twentieth Century-Fox executive Darryl F. Zanuck volunteered to work with the Eisenhower administration to provide "propaganda" messages that would valorize the government in order to help rehabilitate the United States' image in the international arena.[7] This is significant for television studies because it revisits Althusser's argument that the media behave as Ideological State Apparatuses even when privately owned.[8] *Broken Arrow*'s public-relations strategy was to acknowledge crimes committed by the U.S. Army during its conquest of Apache territory in the 1880s but at the same time suggest that the federal government was doing its best to rectify this unfortunate situation and to eliminate racism in the United States. This playing-up-the-positive-aspects strategy followed Lasswell's 1945 recommendation to the U.S. Department of State that it should acknowledge structural problems of racism and then *reframe* its public-relations messages in terms of progress being made in rectifying them. In *Broken Arrow*, the social and judicial systems are depicted as having some problems—that is, certain problems are acknowledged—but these are addressed and the system itself adjusted to the satisfaction of all concerned. This strategy would help improve the government's image during the civil rights movement, which was gaining critical mass, so the American Indian protagonist, in a sense, ventured to speak for all minorities, including African Americans. Yet, ironically, there also appear to be occasions in which Indians were used as a "model minority" in an attempt to discredit claims made by civil rights advocates. This is no surprise, however, since throughout history, Native American representations have been remarkably flexible, often contradictory, and generally useful for purposes that have nothing to do with them or their concerns.[9]

Chapter 4 examined the 1959 series *Law of the Plainsman* (1960), in which the Indian companion metamorphoses into a top-tier lawman. There is a large conceptual leap here: the two characters formerly seen as Friday/Crusoe, Chingachgook/Bumppo, Tonto/Lone Ranger, are fused into a single figure. This semi-superhuman character comprises the best of both races. Even though he is outwardly an Apache, inside he is white.[10] Using his innate "Injun skills" and intuition, he is able to solve cases no white man could ever hope to crack. At the same time he embodies the reasoning capacities of Enlightenment (i.e., Euro-American) Man. The key element here is his education. This thoroughly Anglicized-Indian figure would reappear as Chief Mingo in *Daniel Boone* (1964), in *Hawk* (1966), and in *Nakia* (1974).

Deputy Marshall Sam Buckhart dresses in Euro-American cowboy style, except for his Apache bandanna, which is symbolically equivalent to Superman's "S" emblem. He is awarded all the cinematic devices normally reserved for white-savior figures. Buckhart is an *honorary* white savior. It also appears he may have been an early example of the model-minority trope, used to subtly suggest that if any minorities have "failed," this has nothing do with the U.S. economic system but rather constitutes an individual failing.

Chapter 5 examined the NBC series *Daniel Boone* (1964–1970), which featured actor/singer Ed Ames as Cherokee Chief Mingo, the "half-breed" son of the governor the Virginia colony. As we have seen, Mingo is very much a Friday-type character who assists "pioneer" Boone in the Euro-American conquest of the region (this is historically true for some American Indian groups but not for Cherokees, who actually sided with the British). While Boone is the embodiment of the post–Revolutionary War "New Man," who augments his Euro-American "pluck" with American Indian characteristics, Mingo is a literal hybrid, the embodiment of the best of both races. This trope will reappear in other characters, most notably in *Walker: Texas Ranger* (chapter 8).

Hawk (1966), discussed in chapter 6, carries on the tradition of the super-enforcer of Anglo-American law and order as an Anglicized Indian. The most remarkable aspect of this program is that the Indian character is set in the contemporary era. *Hawk* employs what Gray calls an *assimilationist* discursive strategy in which the minority character is integrated into the dominant (white) society. Indeed, Detective Hawk's Indianness is notably downplayed—nearly eliminated, in fact—in favor of his individual sex appeal and savoir faire: his Indianness emerges only when hypermasculinity, hyperphysicality, or hypersexuality is required for solving a case. Like Superman, Hawk can turn his powers on or off at will. He is also at times depicted as an American James Bond, wearing fashionable and expensive clothes and possessing an aura of confidence that hangs on his respectable, secure, government position that gives him authority, power, and upward mobility. In the final analysis, however, *Hawk* is a rather typical U.S. police drama with a minority protagonist, whose difference from a Euro-American police officer ultimately proves negligible. Yet apparently U.S. viewers were not ready to accept a nonwhite savior figure.

Nakia (1974), discussed in chapter 7, moves to what Gray calls a *multicultural* discursive strategy. This is a synthesis of *pluralist* (separate but ostensibly equal) and *assimilationist*. Sheriff's deputy Nakia Parker is a personification of this dialectic: he tries very hard to reconcile his identity as a Navajo with his position as an apprentice white man, an official representative of the dominant power structure. In the made-for-TV movie that served as the series' pilot, Parker agonizes that he is "caught between two worlds." However, the weekly series turned out to be an altogether different beast.

Here Parker's Indianness fades into the background and becomes almost a nonissue—as if the program's producers simply gave up on the idea of his uniqueness—as he becomes, like Hawk, fully assimilated. This was a regression in terms of American Indian portrayals, which had been steadily growing in moral superiority since Tonto. One explanation could be that the program's producers or network executives decided viewers were not ready for multiculturalism. Indeed, the media's legendary "swing to the right" (i.e., reactionary backlash) may already have been under way at this time.[11] In any event, the series failed; another Indian lawman/protagonist would not appear until nineteen years later with *Walker, Texas Ranger*.

Walker (discussed in chapter 8) is the first successful (i.e., long-running) series featuring an American Indian protagonist—most likely because he is only *half* Indian and generally presents himself as white.[12] As a half-breed, Walker is both cowboy *and* Indian. More importantly, identity for Walker is fluid: he is able to switch back and forth from white to native with little effort, just as Clark Kent is able to become Superman by tearing off his shirt and tie. *Walker* also exemplifies how much all of these protagonists and programs examined herein raise issues of masculinity (women are rarely if ever featured), specifically a uniquely American (i.e., a merged Euro-American and American Indian) version of masculinity, largely built on what had been traditionally perceived as Indian attributes. It is evident that these two positions—white and nonwhite masculinities—work structurally as oppositions. Walker succeeds creating a synthesis of both. Concepts of Euro-American masculinity are extremely fluid and given to change, the foundations are surprisingly stable and predictable. When American ideas of masculinity veer from the traditional macho model, as they did in the 1960s, they usually revert, as they did in the 1980s.[13] Then they soften again for a while, as they did in the 1990s, but then patriarchy strikes back once more. Euro-American masculinity is a complex, ambiguous matter, however, not always perceived in terms of simple machismo.[14] Sometimes it is the Euro-American male's intellect—this is connected to his technology—that designates him as superior. In Walker's case it is, ironically, his knowledge of Asian martial arts. Thus the white savior here is able to beat *both* American Indians and Asians at their own game. Thus the white hero can absorb the attributes of Others he finds valuable: he is a chameleon capable of becoming anything at any time. On the other hand, however, American Indians are generally depicted as either as hypermasculine—that is, savage or berserk—or oddly feminized. There seems to be no rhyme or reason to this alternation except that the depiction employed is whichever one makes the white character look better in comparison.

COLONIALIST FANTASIES

The Indian-enforcer figure (Regulator) is not only a common trope but a recurring one that fits into the schema of colonial discourse, for reasons explained in the introduction. One might reasonably ask, why is a colonial-discourse analysis not employed more often in studies of minorities, especially those of American Indians in film and television? One reason is that leaders and pundits prefer to present the United States as anything but a colonial power. Indeed, the idea of the United States as an imperial power is considered intrinsically wrongheaded: How can this be true when the foundations of the nation were built upon a revolution *against* what was then the world's greatest imperial power?[15] That the United States gradually took over Britain's function as imperialist hegemon in the Western Hemisphere (and elsewhere) is evaded. Whereas the British Empire openly declared itself an imperial power, the United States has put itself in a position, by way of its early anti-imperialist rhetoric, of having to deny this.[16] With few exceptions, Euro-Americans are clearly uncomfortable recognizing the United States as a colonialist enterprise: the United States does not *colonize* other countries, it liberates them. Foreign-policy analyst Andrew Bacevich writes that it is a "cherished American tradition [that] the United States is not and cannot be an empire."[17] Historian Richard W. Van Alstyne has noted, "[I]t is almost heresy to describe the nation as an empire."[18] Media historian J. Fred MacDonald has noted, "[T]he true history of westward expansion is the story of US imperialism."[19]

While media pundits and patriotic series such as *Daniel Boone* reinforce the conventional wisdom that the United States, unlike Britain, has never had imperial ambitions, the fact remains that the U.S. media, including network television, is riven with colonialist discourse, as this study demonstrates. This study has focused primarily on depictions of American Indians in which members of the colonized group are portrayed as *collaborators* or enforcers of the dominant group's norms. This is a typical colonialist fantasy (à la Robinson Crusoe): The subaltern enthusiastically accepts his role as purveyor and enforcer of the dominant group's norms and in doing so abandons his own—and often his own identity as well. The function of this role, I suggest, is a *symbolic endorsement* of the dominant group's norms by the subaltern's symbolic representative. This is, in Churchill's term, a "fantasy of the master race."[20] In other words, the colonist has invented an imaginary subaltern subject that believes in the hegemonic project perhaps more than the colonist himself. Film scholar Robert Stam explains,

> The ideological function of such images is not dissimilar to that pointed out in Barthes's famous analysis of the *Paris Match* cover which shows a black soldier in a French uniform, eyes upraised, saluting what we presume to be the

> French flag. . . . The black soldier's zeal in serving the established law [provides] *the best answer to critics* of that society.[21]

The same can be said of American Indians in TV westerns from *The Lone Ranger* to *Walker, Texas Ranger*. In this case the hegemonic project is called "law and order," which basically means making the territory safe for commerce.

Since representations can be manufactured, reproduced, and disseminated endlessly, it is perhaps even more advantageous—certainly more efficient—to incorporate *representations* of sepoys into the colonizing system rather than physical ones. In any case, the fantasy is more accessible, more malleable, and, ultimately, more valuable than the reality because it can be endlessly replicated and used for a variety of purposes. It follows, then, that televisual Indians are far more valuable for ideological purposes than real ones and that this fantasy film and televisual sphere could be called a *simulacrum*.[22]

STRUCTURE AND OTHERNESS

Why is it that societies often define themselves in opposition to that which they are not? Both Drinnon and Slotkin have demonstrated that both the Pilgrims of Plymouth and the Puritans of Boston cited the "uncivilized" and "heathen" nature of the Pequots as ipso facto evidence that they, the English setters, were by contrast godly.[23] It seems that opposition to the Other is structural in the language—and the thinking—of western civilization (this may be true of all human civilizations) and hence easily activated: perhaps it operates at some base level in the human brain. Structuralism is often dismissed as overly reductive or simplistic, yet it should not be disregarded, because it provides great explanatory power as to why groups of people find it so easy to demonize one another.[24] Euro-American rhetoric (perhaps this is universal) evinces a tendency to reduce complex relationships to simple binaries based on comparisons of self with Other.[25] If "they" are evil, then "we," by comparison, must be good.[26] Structural bias is common in the representational conventions of film and television:

> In the [canonical western's] repeated motif of the encircled wagon, for instance, the camera transforms Native American difference into hierarchical Otherness by locating the primary point of view behind or at the level of the wagon, with sound in the ensuing battle isolating Anglo pain and death and most often collectivizing Indian utterance in the "war whoop." Shot scale and duration—close-ups focusing down on white women and children, for instance, or the wide-angle long shot taking in a horizon filled by hostile figures—wed the technical features of film production to ideologies of race.[27]

Robyn Weigman's statement applies to film, but television westerns (the genre that contains the most American Indian representations) continued mostly in the same directions, using the same visual devices; indeed, most filmed television programs were made (or distributed) by the same studios that dominated the film industry. One big difference, however, was that the U.S. government had far more influence on television than it did on the film industry, since television is regulated.

Both film and TV westerns purport to be based on U.S. history: the past is reframed as a glorious undertaking, the fulfillment of God's wish for his chosen people (i.e., Manifest Destiny), or as a rescue mission designed to rescue the pitiful Other from himself or from some demonic Other. Such fantasies serve to justify and legitimize (i.e., naturalize) colonial norms and practices. It is noteworthy that these tropes remain virtually intact well into the twentieth century. Stam and Spence explain that the colonial enterprise was often presented as a philanthropic "civilizing mission," reframing the colonial presence as a humanitarian intervention rather than an invasion.[28] This claim has more credibility if members of the colonized can be recruited to endorse the invasion or occupation. These can be wholly imaginary. In Orwell's *Burmese Days*, first published in 1934, a disillusioned British imperial policeman (modeled on Orwell himself), cynically asks a local, "How can you make out that we [the British] are in this country for any purpose except to steal?" Enter the Regulator: the native Burmese doctor whom Flory queries entirely supports the British presence. Dr. Veriswami responds that the British colonial effort in Burma entails "a magnificent record of self-sacrifice" on the part of the colonizers.[29] Here we see the subaltern not only endorsing but defending a system that forcibly imposes the norms of an invading culture. We see this same confrontation recur in U.S. television in 1956 with Apache deputy Sam Buckhart's endorsement of U.S. law and social mobility (see chapter 4 on *Law of the Plainsman*).[30]

Racism is a convenient justification, and it usually takes one of two forms: The targeted people are either deemed inherently evil and must be destroyed, as in canonical westerns, or they are helpless objects that must be managed, as in the case of James Cameron's *Avatar*. In either case, the targeted group usually has something the colonizing group wants (in *Avatar* it was the mineral unobtainium). This aspect was clearly addressed in 1848 by Albert Gallatin, a Swiss immigrant who had been secretary of the treasury under presidents Jefferson and Madison. Gallatin noted that racist rhetoric was a smokescreen for greed: "The allegations of superiority of race and destiny . . . are but pretenses under which to disguise ambition [and] cupidity."[31] In the case of American Indian lands, the resources—besides the land itself—included timber, gold, copper, uranium, and water. Sometimes the desire for the Other's resources creates a *retroactive* necessity to demonize the Other, fomenting hatred where there may previously had been none.

Religious and ethnic hostilities can be exploited, or, in their absence, invented.[32] Memmi asserts that racism is often a retroactive justification for an already existing—or planned—colonization, with hatred coming "in the wake" of actual oppression.[33] Stam further elucidates this point: "American Indians were called beasts *because* white Europeans were slaughtering them and expropriating their land"—not the other way around.[34] Thus it became necessary to create literary and filmic representations to *retroactively* justify taking their land. It is crucial to note that this retroactive justification was still being undertaken in movies and television *one hundred years after the fact.* Actor John Wayne, the personification of Anglo-American masculinity and cowboy culture, outlined his own thinking along these lines in 1971:

> I don't feel we did wrong in taking this great country away from them [Native Americans]. . . . Our so-called stealing of this country from them was just a matter of survival. There were great numbers of people who needed new land, and the Indians were selfishly trying to keep it for themselves.[35]

According to Wayne's reasoning, American Indians *deserved* to be decimated and forcibly relocated because *they were selfish*, whereas, it is structurally implied, Euro-Americans were noble, concerned for the welfare of the many rather than the few. The fact of the matter was that most Native Americans saw the land as a commons and might have been satisfied with sharing it with Euro-Americans had they not discovered that Euro-Americans planned to exclude them and selfishly hoard the land for *themselves*.[36]

FANTASIES, WISHFUL THINKING, AND REASSURANCE

Rather than being consciously intended to degrade or demoralize American Indians, as Clark suggests, the representations I have analyzed more likely indicate wishful thinking (i.e., fantasies) on the part of the dominant group. In other words, they provide *reassurance* for Euro-American viewers, who comprised the overwhelming majority of the audience. *New York Times* columnist Maureen Dowd noted that stereotypes are above all *comforting*: "They wrap life in the archetypal toastiness of fairy tale and myth. They make complicated understandings unnecessary. They exempt us from further mental or emotional effort."[37] Film scholar Robin Wood notes that such fantasies "construct the adult spectator as a child" seeking reassurance of his or her own innocence.[38] This comforting-images concept is a still-emerging perspective that should be developed further.[39]

A strong argument could be made that racist (and sexist) fantasies provide ideological gratification for viewers of the dominant group.[40] French sociologist Pierre Bourdieu explains that pleasure occurs when the viewer agrees with the ideology purveyed in the texts because the viewer is *reassured* that he or she has assumed the correct position.[41] Citing Kant, Bourdieu asserts that "no judgment of taste is innocent"; this suggests that the viewer/spectator is implicated in the process and *actively seeks the desired outcome* to the exclusion of all others.[42] Barthes delved into the complexities of this process in 1973. He notes that part of the pleasure of interpreting (fictional?) texts is in seeing the dominant ideology *challenged and then restored.*[43] In fact, the ideological system is left in better condition than it was previously, due to the fact that the challenge has been absorbed (or co-opted) by the system itself. Media scholar John Fiske explains, "[T]he restorative event is necessarily performed by the sociocentral hero, who restores the status quo—or preferably an enhanced version of it."[44] This is clearly the case in police programs such as *Nakia*, in which the protagonist, a minority figure, undergoes challenges with regard to his commitment to the dominant ideology yet emerges with his faith in this ideology restored; in this process, the norms of the dominant group are not only reaffirmed but strengthened. Fiske concludes that gratification is "the product of accommodation to the dominant ideology."[45]

If readers/viewers obtain ideological satisfaction, are they themselves then guilty of propagating stereotypes? According to Fiske, yes: "[W]e [viewers] are indulging in an ideological practice ourselves; we are maintaining and legitimating the dominant ideology, and *our reward* for this is the easy pleasure of the recognition of the familiar."[46] Homi K. Bhabha adds that the viewing or reading experience, in order to be most effective, should be both terrifying *and* gratifying, something akin to watching a horror movie.[47] Bhabha addresses the pleasurable aspect of the viewer's being on the "correct side of ideology" in that this process flatters the viewer, elevating him or her to a position of superior being; he describes this pleasure as narcissistic.[48] Certain basic visual strategies do appeal to—or encourage—narcissism: for example, the all-seeing, omnipresent camera (discussed in chapter 4 on *Law of the Plainsman* and in chapter 6 on *Hawk*) often positions—or "constructs"—the viewer as God.[49] This implies that the viewer's judgment is infallible and, as Bourdieu points out, places the viewer "on the correct side" of whatever ideology is enunciated in the text, assuming he or she acquiesces to the rhetoric.[50] In other words, this is not merely entertainment; there appears to be a sort of a negotiation or quid pro quo at work here: the viewer receives a story and in return is expected to acquiesce to the story's ideology. This implied contract indicates persuasion is at work, not mere storytelling: a certain point of view is being "sold." The viewer/reader is flattered if the ideology of the text happens to agree—or gives expression to—his own beliefs, which are thereupon validated. Claire Johnston writes that this operation bears "a particular function in relation to paternal ideology."[51] Indeed, most westerns and crime dramas (the genres that contain the most American

Indian characters) are inherently male oriented and authoritarian, a fact clearly indicated by their dearth of female characters.

There are other aspects that appeal to viewers' narcissistic proclivities. At an even more basic level, the fact that a fiction narrative appears to be organized for—and around—the reader/viewer implies that he or she sits at the center of the diegesis (with the camera as a "stand-in" or surrogate); the fictive universe more or less revolves around the reader/viewer.[52] This "primary identification" provides the reader/viewer with a feeling of power.[53] It must be noted, however, that these theories of identification are based on Lacanian psychoanalytic theory, which is assumed to apply more to film than to television because the viewing experience in a movie theater is different from that experienced at home.[54] Some scholars assert that primary identification (with the camera) is drastically curtailed in the home due to the distracted nature of the television viewer, who is not immersed in the viewing experience in the way a moviegoer would be.[55] However, researcher Pieter Fourie insists that both primary and secondary identification are present in television viewing and accessible to the degree that the viewer is willing to participate.[56] "Secondary identification" means that (most) viewers search for a character to empathize with, generally the protagonist.[57] As chapter 2 on *The Lone Ranger*, chapter 3 on *Broken Arrow* (with respect to Tom Jeffords), and chapter 4 on *Law of the Plainsman* demonstrated, many if not all the protagonists in these programs are presented as gods themselves or as saviors sent by God. There are several visual strategies at work in these programs, which flatter viewers and construct them, too, as divine. As Richard Dyer extensively demonstrates, European (and by extension Euro-American) art, including film (and by extension television), has for centuries worked to construct white people—especially the presumed white male subject—as godlike.[58] This appears to be a cultural imperative, and it is apparent in most if not all of the programs analyzed in this study. Granted, the American Indian protagonists in these programs (the characters, not the actors) are not officially "white," but they are on the *side* of the whites; hence they function as *honorary* white men. What could be more flattering to a white male viewer than watching a minority character trying to be like him when he (the viewer) can be white (and male) without any effort? It is useful, then, to reconsider Bhabha's remark in this light: it is quite possible that as an advertising medium, U.S. television actively appeals to viewers' narcissistic tendencies in an effort to flatter them and create a "positive affect"—in this case, a feeling of moral superiority. As noted earlier, advertisers hope viewers will associate—or "attach"—this flattering, gratifying feeling to the product being advertised.[59]

It might be illuminating to consider who the audiences for these programs are and what they tend to expect. As we have seen, programs featuring American Indians largely fall into the western genre, although, as this study demonstrates, sometimes police dramas set in the contemporary era, such as *Hawk*, *Nakia*, and *Walker*, incorporate Indians as well. What is more, there seems to be a close kinship between these two genres (and their audiences), and in some cases the two are combined as hybrid forms. Who are the audiences for such programs, and what do they like to see? Surely it is in the advertisers', producers', and networks' best interests to provide them with what they expect, whatever that might be. According to scholar Yardena Rand, the audience for westerns (she refers to both film and TV westerns) mostly consists of white, middle-aged, undereducated males:

- 79 percent white
- 63 percent male
- 71 percent forty-five years or older
- 76 percent have less than a bachelor's degree
- 63 percent earn less than fifty thousand dollars per year
- The largest audience bloc lives in the South.[60]

The western, Rand adds, "is hardly known for being socially progressive":

> The heroes of westerns are almost always white men, and it is *their* struggle, crises, decisions and actions that form the foundation of the genre. White women and people of color function primarily as ancillary characters, necessary presences to define the hero and further the plot, but not [as] complete in and of themselves—women to protect or dominate; Indians to kill or serve as helpers; blacks or Asians, if we see them at all, as servants.[61]

Richard Abel goes so far as to describe westerns as "white-supremacist entertainment."[62] It is not surprising that many of the predominantly white male viewers of these programs identify with these heroes and operate vicariously through them, in a sort of simulacrum. Indeed, several of Rand's interviewees confirm this: "I was always the cowboy"; "I want to be the good guy"; "I can become a star and participate just by feelings. . . . I seem to become part of the play."[63] From Rand's profile (see earlier list), it seems reasonable to infer that audience members for westerns are generally working class; one could describe them as mostly "blue collar," including some lower-scale "white collar" types. English professor and social critic Paul Fussell asserts that this group on the whole has an identifiable set of proclivities when it comes to entertainment: "The wholly predictable is what [proles] want."[64] Again, we see that typical viewers for such programs do not wish to have their conventional views—nor their prejudices—challenged. In fact, they are likely expecting their views to be reinforced and enjoy seeing the status quo validated. It follows that this would be what advertisers wish to give them. Fussell cites sociologist Arthur B. Shostak: "[Blue-collar types] choose *leisure pursuits that have the power to affirm acquired wisdom* rather than provide any confrontation with novel and [mentally] taxing matters. They choose leisure pursuits that have an ability to 'massage' rather than for any message the medium might contain."[65] This would be in line with Cummings's concept of "comforting images." Vine Deloria Jr. remarked, "[Westerns] seem to rely on a few dramatic story lines that are *predictable*; consequently *westerns often reinforce conservative fantasies*."[66]

Much the same could be said for crime dramas—indeed, it could be argued that the western itself often takes the shape and form of a crime drama (when it is not a war drama). In fact, there are several series examined in this study that actively blur the boundaries—if indeed there are any—between western and crime drama, such as *The Lone Ranger*, *Law of the Plainsman*, *Nakia,* and *Walker, Texas Ranger*. Like westerns, crime dramas offer ideological and psychic reassurance for conservative viewers, some of whom are anxious about social change. Media professor Aurora Wallace of New York University said, "[With police dramas] you've got basically a fight of good against evil. . . . I think what the audience is looking for is reassurance."[67] Author and scholar Jason Mittell writes, "Police dramas [are] conservative *rituals of reassurance*."[68] As discussed in chapters 6 and 7, (on *Hawk* and *Nakia*, respectively), we may surmise that the ostensibly conservative audiences for both westerns and crime dramas corresponded with the group Richard Nixon dubbed his "silent majority."

TELEVISION WATCHING AS RITUAL

Rituals are regularly occurring, repetitive, communal activities usually designed to provide reassurance, to help assuage anxieties for individual members of a culture as well as to provide a bonding experience to help create a collective identity.[69] Nachbar and Lause describe moviegoing as having an especially ritualistic quality during the Great Depression.[70] Television watching also incorporates ritualistic behavior in that regular programs are seen at the same time every week.[71] Moreover, certain categories of films and TV programs, such as westerns, generally follow patterns that are strictly formulaic: these are often so predictable they could be considered ritualistic and therefore *reassuring*. Thomas Schatz calls the western an American "foundation ritual."[72] The main thing viewers of westerns can be reassured of is that the righteous white-male individual will prevail over most if not all challengers. This comforting aspect might help explain the western's vast popularity and appeal during the civil rights era, when white patriarchy and privilege was being questioned.

Much the same could be said of the police/crime drama, which like the western is often so predictable as to be ritualistic. Again, with few notable exceptions, the paternalistic, righteous white male prevails. These programs can provide a sort of compensatory fantasy or dreamlike denial of the challenges to white-male privilege that encroach in the real world. Hence, one might expect this type of program to exert a stronger appeal during periods when patriarchy is being questioned, as it was by feminism in the 1990s when *Walker, Texas Ranger* became a top-rated series. According to Bernadette Casey and colleagues, television-genre theory "focused on ways in which generically defined structures can help to build and maintain certain (conservative, patriarchal) ideologies and values which act as a form of reassurance to the audience."[73]

If the crime series is a *celebration* of white patriarchy, then the highest honor it could confer would be welcoming women and minority characters into this position (i.e., as Regulators). Nonetheless, female and minority characters constitute a small proportion of authority figures seen on TV. Moreover, these characters are screened carefully—and their loyalties tested thoroughly—before being welcomed into the "boys' club": they usually serve as low-level functionaries who report to white-male superiors. Although a few "token" individuals are accepted into the (lower) ranks, the system itself remains basically unchanged. Thus the threat of minority and female encroachment in the arena of white patriarchy is *contained*—tightly controlled and authorized by white males themselves, who also happen to comprise the vast majority of television's producers and network executives.

COGNITIVE DISSONANCE

Another important aspect of this study, one several scholars have noted, is that American Indians are most likely to be depicted in a fictional-historical past, as if they and their current issues do not exist—or no longer matter. This appears to be part of the psychological strategy, a defense mechanism, for dealing with cognitive dissonance that Clark/Khatib calls "temporal displacement" (discussed in more detail in chapter 3 on *Broken Arrow*): issues that might create discomfort for viewers can be better dealt with by placing them somewhere else in time; it could be the late 1800s, as is the case in most film and television depictions of Indians, or in the future, as in science fiction (*Star Trek*, *Bravestarr*, or *Avatar*). In any case, the issue the dominant culture is most likely struggling with is guilt over its genocidal policies: by displacing these onto another era, the issue is "retemporalized," making it appear distant or even irrelevant, ameliorating the viewer's discomfort.[74] Stephanie Greco Larson explains,

> Revisionist [i.e., post–World War II] westerns relegate racism to history by admitting that something bad happened long ago. These stories acknowledge *some* of what was done against Indians during Western expansion and provide

> sympathetic representations of Native Americans. By acknowledging this, *the white audience gets to feel absolved of guilt* for what [its] predecessors did.[75]

Thus, setting American Indians in another era lets viewers "off the hook" because, ostensibly, nothing can be done at this point—it is too late to worry about it now.

COLLECTIVE TRAUMA, RITUAL EXORCISM, AND SYMBOLIC EXPIATION

What is the role of the viewer in this discourse? Thomas Schatz, as structuralist, suggests that there is no clear dividing line in the exchange between producer and viewer: both parties participate actively, and this transaction represents a "ritualization of collective ideals, the celebration of temporarily resolved social and cultural conflicts, and *the concealment of disturbing cultural conflicts* behind the guise of entertainment."[76] This poses the interesting implication that there might be some kind of collective *psychic trauma* that U.S. culture is attempting to deal with on a symbolic plane.[77] This seems rather a priestly function; indeed, Gerbner has asserted that television has assumed some of the functions of religion.[78]

There must have been a sense of collective shame that needed to be dealt with—or denied—during the civil rights movement. American Indians were used as stand-ins for other minorities—or for minorities in general—as a way of implying, "as you see, we treat our minorities fairly in this country, despite what the rest of the world might say about us." This supports Schatz's assertion that this type of discourse was not mere entertainment. There must have been some shame about the United States' crimes against American Indians that was—and perhaps still is—being "worked through" in films and television: for example, when asked by Peter Bogdanovich whether his film *Cheyenne Autumn* (1964) had been an apology of a sort to American Indians, John Ford replied, "Let's face it, we've treated them very badly—it's a blot on our shield: we've cheated and robbed, killed, murdered, massacred and everything else."[79] Certainly there must be some collective guilt felt at some level—at least by some segment of the citizenry—when a society's atrocities are laid bare, as they were, for example, during the Vietnam conflict. Can it be mere coincidence that the rise of the television western in the 1950s correlated with the conflict in Vietnam and that both peaked in the late 1960s? It was only a matter of time before someone made the connection between the Sand Creek massacre of 1864 and the massacre at My Lai in Vietnam in 1968. Both were painful episodes that needed to be worked through. Historian Hayden White describes these as "holocaustal events": "They cannot be simply forgotten and put out of mind, [nor] *can they be adequately remembered*."[80] Mark Golub writes, "What is at stake is how to represent events that are, by their nature, too traumatic to be adequately depicted."[81] Arthur Penn used *Little Big Man* (1970) to deal with—or expiate—a sense of national shame attached to My Lai by comparing it to Sand Creek; *Soldier Blue* (also 1970) draws the same comparison.[82] Regarding this spate of guilt-ridden films, Donald L. Kaufman remarks, "The white psyche, after over two centuries of unacknowledged racism, finally succumbed to an overdose of remorse."[83] Thus these inner conflicts must be worked through on film and on television.[84] Although psychoanalytical methods are beyond the scope of this study, it is worth considering whether an entire society and its cultural products can be psychoanalyzed.

There are several methods Euro-Americans regularly use to deal with—or deny—their collective trauma. Negative stereotyping seems to have been the earliest and crudest: it blames victims for their own misfortune. Another, slightly more sophisticated, method is justification. According to psychologist John Sabini, justification works by claiming that whatever transgressions were committed were in fact reasonable reactions, such as claiming that the other party attacked first and therefore violence was a reasonable response in terms of self-defense.[85] "Projection," a term coined by Freud, is an *advanced* form of justification wherein the belligerent party attributes its *own* hostile intentions to its victims, thereby justifying a preemptive strike against an attack that has not yet materialized (whether it ever materializes is beside the point).[86] Drinnon, Slotkin, and Berkhofer have all suggested that the New England Puritans, in their demonization of the Pequots, were *projecting* their own hostile intentions—as well as their repressed sexual desires—upon the aboriginals.[87] Linguist and social critic Noam Chomsky explains,

> It turns out that the groups who are about to destroy us are the ones that we're destroying. So in the earliest period [of U.S. history], the Indians were coming to destroy us. In the first major executive war, the conquest of Florida [the Seminole Wars led by General Andrew Jackson], we were defending ourselves from runaway slaves and lawless Indians who were going to destroy us while we were exterminating them.[88]

Aggressive impulses must be denied even if denial involves bizarre contortions of logic. Many if not most Euro-Americans do not want to confront the fact that the winning of the West was primarily a process of forcible colonization.[89] Several scholars have suggested that U.S. television has elided these issues and enabled Euro-Americans to evade guilt by offering "comforting images" that downplay rather than deny wrongdoing.[90] Raul Chavez writes, "Colonizer literature and cinema exists because Americans are comfortable with their soothing message of social and cultural (and religious) righteousness."[91] Author and cultural critic James Baldwin noted how U.S. culture spends a great deal of psychic energy creating images whose purpose is to enable Euro-Americans to evade taking stock of crimes their nation has committed against nonwhites: extermination of Native Americans, enslavement and suppression of African Americans, illegal imprisonment of Japanese Americans during World War II. These efforts at evasion damage the nation's psyche, Baldwin explains: "This failure to look reality in the face diminishes the nation as it diminishes a person."[92] In the documentary film *Images of Indians*, American Indian Movement cofounder Russell Means agrees: "This country refuses to face what happened to Indian people. This allows the American people to remain ignorant—and to repeat history. . . . We [American Indians] have a saying: a nation that does not know its own history has no future."[93] Explaining these efforts in the context of strategies of denial, Stephanie Greco Larson offers, "Since the reality of United States' history of genocide against indigenous people is difficult to accept in a modern context, films and entertainment television that include [depictions of] Native Americans serve to deny, justify, or forgive the atrocities committed against them."[94] Yet even on the rare occasions when Hollywood has admitted that crimes against humanity were committed, as in the film (1950) and television versions of *Broken Arrow* (1956–1958), these accusations are deflected: they are attributed to a few sociopathic individuals rather than being systemic. Thus the system is recuperated while a few "bad apples" are symbolically ostracized.

Whereas earlier, cruder justifications operated by portraying the extermination of subhumans as a moral or Malthusian imperative or as a matter of self-defense, post–World War II (i.e., revisionist) westerns have operated in a more subtle, sophisticated manner, designed to appeal to Euro-Americans' humanitarian instincts by portraying their government as having *protected* American Indians rather than systematically suppressed them or allowed corporations to colonize their land and resources.[95] Working together, Hollywood and the television networks obscured the fact that the colonization of native land and blatant disregard for American Indians' health and safety was still ongoing during the Cold War—and is ongoing even today.[96]

U.S. television avoids American Indian issues because most advertisers want viewers to associate their products with feel-good images: the last thing advertisers want is for their products to be juxtaposed with images that create discomfort, anxiety, or guilt. Annette M. Taylor explains,

> Popular culture generally avoids contemporary, real-life Native Americans and Native American issues because of *national collective guilt* about government policies of genocide. Guilt makes people feel bad, and *people who feel bad do not feel like spending money*. And the purpose of [commercial] television, after all, is to sell products.[97]

A frequent element of denial in these movies and television series is that injustices are committed by deviants and renegades, yet the system itself is righteous. In *Broken Arrow* (1950), *Dances with Wolves* (1990), and *Avatar* (2009), the message seems to be that, yes, there were *some* bad Euro-Americans who committed atrocities, but the true heart of the American people—represented by the white protagonist who fights the evil whites on behalf of the natives—is noble and just. In the words of Ward Churchill, such efforts constitute "a glib contrivance designed by and for the conquerors to promote their sense of psychic reconciliation."[98]

MORE ON NARCISSISM

Ralph and Natasha Friar note, "By explicitly justifying the genocide perpetrated by our forefathers, Hollywood utilizes our ignorance to [reinforce] our egotism."[99] The word "narcissism" could well replace "egotism" in this instance. Todorov also suggests that egocentrism is what is ultimately at stake: these "figures of the experience of alterity are grounded in egocentrism, in the identification of our own values with values in general, of our *I* with the universe."[100] In other words, it is only right and natural that we deserve and should have what we want because we are naturally superior; we are God's "elect."

By 1951, thanks to network television, Euro-Americans could satisfy their fantasy addiction, which had heretofore been indulged in movie theaters, in the comfort of their homes. U.S. television became a simulacrum, a fantasy substitute for reality, very much like Walt Disney's Frontierland.[101] This allowed Euro-Americans to enter a dream world where they rarely if ever did anything wrong to anyone and could continue to see themselves as good Christians. Television flattered Euro-Americans, telling them what they wanted to hear: that they were a noble, innocent, and altruistic people, concerned only with helping the less fortunate—the modern version of Kipling's "white man's burden." Enter the Regulator: he is here to reassure the Euro-American public that American Indians do not mind after all, that in the long run the Anglo-American way was better for all concerned, even if many (if not most) Indians had to die. This need became particularly poignant during and after World War II, when Euro-Americans became horrified by what the Nazis had been doing to minorities in Europe.[102] Euro-Americans did not want to be reminded (by the Soviets) that they themselves had committed similar crimes.

Chirot and McCauley explain, "[Euro-]Americans reduced the dissonance between their professed ideals and the reality of the immense suffering they imposed on Native Americans by claiming it was 'for their own good.'"[103] According to the colonialist fantasy, some of the natives—a select few—were astute enough to recognize that the Euro-American position was both superior and desirable.[104] This was plainly depicted in the series *Brave Eagle*

(1955). The collaborator not only endorses Euro-American conquest; he reassures Euro-Americans that their conquest represents the greatest good for the greatest number. The following lines come from Sam Buckhart of *Law of the Plainsman*. Buckhart intones,

> At one time my highest ambition in life was to be the bravest warrior in the Apache nation. My people were at war. It was a futile war, and the Apache lost. Some of my people held bitter resentments. But there were others who saw *the great opportunities* that lay ahead, and they took advantage of them—because they wanted to.[105]

Buckhart wants a piece of American pie. He decides it is better to be a second-rate citizen, a junior partner, in Anglo-American society than a full citizen in Apache society. He turns his back on the Apaches' collectivist ways, adopting instead the economic individualism of Anglo-Saxon America. He abandons stifling collectivism, preferring the every-man-for-himself approach to upward mobility. Underneath this lies a Cold War implication that American Indians, left to their own devices, are communists at heart.[106] Also like communists, they constitute an obstacle to the spread of Anglo-Saxon capital.

AMERICAN IMPERIAL MYTH, THEN AND NOW

The restless search for profit-making and the removal of anything—or anyone—standing in the way of doing so have always been central to the American character, writes foreign-policy analyst Andrew Bacevich: "From the outset, Americans evinced a compulsion to acquire territory and extend their commercial reach."[107] Given the Puritan underpinnings of U.S. culture and philosophy, this should come as no surprise. Plymouth Colony, founded in 1620 by a radical sect of Puritan separatists (the so-called Pilgrims) who wanted to split off entirely from the Anglican church, was an experiment in theocratic capitalism—as well as one of the earliest corporations.[108] Its objective was to establish a theocratic, divinely sanctioned state, modeled on Calvin's Geneva, in which Protestants would work hard, invest, expand (i.e., colonize adjoining areas), and grow rich, as God intended. The more moderate Puritans who founded Massachusetts Bay Colony in 1628, led by John Winthrop, also considered themselves Christian businessmen and that their project, too, was God's work.[109] Winthrop and his followers conceived this "new land," like Calvin's Geneva, as a fusion of theocratic imperium and commercial enterprise, one that would expand, eventually encircling the entire globe. Horsman writes, "[T]he Puritans saw not only God's kingdom moving to the West but also thought of America as the place from which the renovation *of the world* would begin."[110]

The western tale, as several authors have noted, is replete with Puritan themes. The underlying message (or meta-narrative) of the western, as a genre, is that the juggernaut of Anglo-American capitalism is both desirable and inevitable; hence the term "Manifest Destiny." Obstructions to commerce, whether geographical (mountains, rivers, deserts, etc.) or human, were to be surmounted at all costs. Human obstacles appeared as Indians or marauding gangsters ("outlaws"); in either case, such threats had to be contained so commerce could continue to thrive and expand.[111] It was easy to transfer this Puritan ethos to other genres: the western, the gangster film (set in the "urban frontier"), anti-Nazi or anti-Japanese World War II movies, anticommunist films of the 1950s: all these implicitly deal with threats to the continuing expansion of Anglo-Saxon capitalism.[112]

RAW MATERIALS AND RESOURCES

There is little question that raw materials and resources—including the land itself—are key drivers of colonialism, as British colonialist Cecil Rhodes explains.[113] This was also true in the United States during the winning of the West, although it is rarely acknowledged explicitly, usually hidden under layer upon layer of ideology. Coincidentally or not, the doctrine of Manifest Destiny took shape at the same time a need developed for industrial minerals, especially copper, which comprised—and would remain—a crucial element in the then-burgeoning communications industry. Copper, for example, was a primary issue in the displacement of the Apaches in New Mexico (see discussion in chapter 3 on *Broken Arrow*). During the Cold War, access to uranium—much of it located on Native American lands in the Southwest—was a key, although oft-elided, issue.[114]

It is illuminating to put this search for resources in a worldwide context. What happened to American Indians on this continent is hardly unique; in fact, one point of this study has been to situate their struggle among other peoples of the planet whose land, resources, and labor have been colonized. Anywhere there are valuable resources, one can find an industrialized nation attempting to control them. The relationship between the control of industrial resources and colonization was recognized in the 2009 movie *Avatar*, which characterized the conquest of the Na'vi people and their planet as a quest for control of the extremely rare and valuable mineral unobtainum, which, like uranium (or even petroleum), comprises a crucial "strategic defense resource." As long as there are "developed nations" who need certain resources to fuel their economies (and war machines), and as long as there are "undeveloped" nations sitting on those resources, there will be wars to gain control of those resources. There will also be colonial discourse designed to confound the issue and cover cupidity in a smokescreen of ideology. This works best when the colonizers can convince *themselves* they are doing the colonized a favor. The overarching theme of those American Indian depictions examined this study—regulators, collaborators, enforcers of the dominant group's norms—is the same as that of *Paris Match*'s black soldier saluting the tri-color: it is the self-congratulating fantasy that colonized peoples (or subaltern minorities) are delighted to become junior partners in the enterprise of their own subjection.

NOTES

1. Grossberg in John Storey, ed., 1998, 617.
2. Also see Tzvetan Todorov, *The Conquest of America: The Question of the Other* (New York: Harper and Row, 1984), 147.
3. Philip J. Deloria, *Playing Indian* (New Haven, CT: Yale University Press, 1998), 10, 22.
4. Ibid., 3.
5. Stuart Hall, "The Question of Cultural Identity." In Stuart Hall, David Held, and Tony McGrew, eds. *Modernity and Its Future* (Cambridge: Polity Press, 1992), 274–316.
6. Christopher Lasch, *The Culture of Narcissism: American Life in an Age of Diminishing Expectations* (New York: W. W. Norton, 1979), 115; also see Todorov, 1984, 165–67.
7. Jackson Committee, "March Meeting Notes." April 10, 1953. Washburn Abbott Papers, Box 7. Dwight D. Eisenhower Library, Abilene, Kansas.
8. Althusser, 1971, 154. This is not always the case, however. The television industry, even though it is regulated, has had its battles with the federal government—note the outcome with regard to Nixon's "enemies list" in 1974—but generally television networks tend to behave as "adjuncts of government." Noam Chomsky, *Necessary Illusions: Thought Control in Democratic Societies* (Boston: South End Press, 1989), 75; also see James Aaronson, *The Press and the Cold War* (New York: New York University Press, 1990), 226.
9. Philip Deloria, 1998, 3, 129, 185.
10. Some American Indians have a pejorative slang term for a person who is "red on the outside, white on the inside": an apple.
11. Director George Lucas is often credited with launching a return to militaristic themes with *Star Wars* in 1977. See Michael Ryan and Douglas Kellner, *Camera Politica: The Politics and Ideology of Contemporary Hollywood Film* (Bloomington: Indiana University Press, 1998), 228–30; also see Robin Wood, *Hollywood from Vietnam to Reagan* (New York: Columbia University Press, 1986), 174.
12. Walker, however, was not the only half-Indian protagonist on television. As noted earlier, the precursor is Hondo Lane of *Hondo* (1967).
13. It seems quite apparent that the Beatles, the Rolling Stones, and other musical groups of the 1960s "British Invasion" succeeded in challenging and redefining masculinity for American youth, as did the San Francisco–based hippie movement. However, such deviations were successfully "rolled back" in the late 1970s and throughout the 1980s, when the Cold War militaristic paradigm returned; see Wood, 1986, 164.
14. Martin Pumphrey, "Why Do Cowboys Wear Hats in the Bath? Style Politics for the Older Man." *Critical Quarterly* 31, October 1989, 78–100. Reprinted in Douglas Pye and Ian Cameron, eds. *The Movie Book of the Western* (London: Cassel, 1996), 51–54.
15. Andrew Bacevich, *American Empire: The Realities and Consequences of U.S. Diplomacy* (Cambridge, MA: Harvard University Press, 2002), 218; also see Stanley Lens, *The Forging of the American Empire: From the Revolution to Vietnam, a History of U.S. Imperialism* (London: Pluto Press, 2003), vii, 39.
16. Loring Wirbel, "Star Wars: Imperialism in Space." In Carl Boggs, ed. *Masters of War: Militarism and Blowback in the Era of American Empire* (London: Routledge, 2003), 77. Also see Lens, 2003, vii, 1, 39; Edward Said, *Culture and Imperialism* (New York: Vintage Books, 1993), 242. Certain government insiders, such as Norman Davis, Franklin D. Roosevelt's

ambassador to the Vatican, recognized the United States' emerging role as imperial power and world hegemon. In October 1940, thirteen months before the Japanese attack on Pearl Harbor, Davis told Jay Pierrepont Moffat, U.S. Ambassador to Canada, that the United States would assume the maintenance of the British Empire after the war: "We shall be in effect the heirs to the [British] empire, and it is up to us to preserve its vital parts." Jay Pierrepont Moffat, Diary entry, October 10, 1940. In Nancy H. Hooker, ed. *The Moffat Papers: Selections from the Diplomatic Journals of Jay Pierrepont Moffat, 1919–1943* (Cambridge, MA: Harvard University Press, 1956), 333.

17. Bacevich, 2002, 218–19.

18. Richard W. Van Alstyne, *The Rising American Empire* (New York: W. W. Norton, 1960), 6.

19. J. Fred MacDonald, *Who Shot the Sheriff? The Rise and Fall of the Television Western* (Westport, CT: Praeger, 1987), 112.

20. Ward Churchill, *Fantasies of the Master Race: Literature, Cinema, and the Colonization of American Indians* (San Francisco: City Lights Books, 1998).

21. Robert Stam and Louise Spence, "Colonialism, Racism, and Representation." In Bill Nichols, ed. *Movies and Methods: An Anthology*. Volume 2 (Berkeley: University of California Press, 1985), 639; emphasis added.

22. Jean Baudrillard, *Simulacra and Simulations* (New York: Semiotext(e), 1983). Beaudrillard offers Disneyland as a primary example of a simulacrum, but films and television are closely related and in fact are Disney's bailiwick.

23. Richard Drinnon, *Facing West: The Metaphysics of Indian-Hating and Empire-Building* (Minneapolis: University of Minnesota Press, 1980), 55; Richard Slotkin, 1973, 18, 25–55.

24. On recent views of structuralism see Stephen Crofts, "Authorship and Hollywood." In John Hill and Pamela Church Gibson, eds. *The Oxford Guide to Film Studies* (Oxford: Oxford University Press, 1998), 316–18; on the continuing validity of structuralism see Todorov, 1984, 153.

25. William Greider, "Acceptable Villains Make Our Troubles So Manageable." *Washington Post*, July 15, 1979, E1, 4.

26. Jeff McArn, "Reflections on Political, Religious Uses of the Term 'Evil.'" *Observer-Dispatch* (Utica, NY), June 13, 2004. http://collegenews.org/editorials/1980/reflections-on-political-religious-uses-of-the-term-evil.html.

27. Robyn Wiegman, "Race, Ethnicity, and Film." In Hill and Church Gibson, 1998, 165.

28. Stam and Spence, 1985, 636.

29. George Orwell, *Burmese Days* (New York: Mariner Books, 1974), 50–51.

30. *Law of the Plainsman*, episode 28, "Jeb's Daughter."

31. Quoted in Reginald Horsman, *Race and Imperialism* (Cambridge, MA: Harvard University Press, 1981), 271.

32. Christopher Simpson, *Science of Coercion: Communication Research and Psychological Warfare, 1945–1960* (Oxford: Oxford University Press, 1994), 94–106.

33. Albert Memmi, *The Colonizer and the Colonized* (New York: Orion Press, 1991), 186.

34. Stam and Spence, 1985, 636; emphasis in original.

35. Richard Warren Lewis, "Candid Conversation: John Wayne." *Playboy* 209, May 1971, 78. Reprinted in Barry G. Golson, ed. *The Playboy Interview*. New York: Playboy Press, 1981, 269.

36. Vine Deloria Jr., *Custer Died for Your Sins: An Indian Manifesto* (Norman: University of Oklahoma Press, 1988), 30, 51, 178.

37. Maureen Dowd, "Cuomos vs. Sopranos." *New York Times*, April 22, 2001, WK17.

38. Wood, 1986, 163.

39. Also see Mark Crispin Miller, "Deride and Conquer." In Todd Gitlin, ed. *Watching Television* (New York: Pantheon Books, 1986), 213. The concept of audience reassurance is often dealt with in reception studies, especially with regard to genre theory, sometimes under "uses and gratifications" in communications science.

40. Wood, 1986, 162–66.

41. Pierre Bourdieu, *Distinction: A Social Critique of the Judgment of Taste* (Cambridge, MA: Harvard University Press, 1984), 152, 214, 598.

42. Ibid., 622.

43. Roland Barthes, *The Pleasure of the Text* (New York: Hill and Wang, 1975; originally published 1973), 14. French sociologist Pierre Bourdieu comes to a similar conclusion in *Distinction: A Social Critique of the Judgment of Taste* (Cambridge, MA: Harvard University Press, 1984; originally published 1979), 152, 214, 598.

44. John Fiske, *Television Culture* (London: Routledge 1989), 112–13.

45. Ibid., 225.

46. Ibid., 12; emphasis added.

47. Homi K. Bhabha, "The 'Other' Question: Difference, Discrimination, and the Discourse of Colonialism." In Russell Ferguson, Russell, Martha Gever, Trinh T. Minh-ha, and Cornel West (eds.), *Out There: Marginalization and Contemporary Cultures* (Boston: MIT Press, 1990; originally published in Screen, 24[6], 1983), 71–87.

48. Ibid., 81–83. Daniel Hallin also points out that television appeals to viewers by flattering them in various ways. "We Keep America on Top of the World." In Gitlin, 1986, 16, 38.

49. Aumont et al., 1992, 214–15. These flattering "omniscient" visual techniques, however, must occasionally be countered by strategies in which narrative information is withheld or restricted in which the viewer knows very little: if the camera/viewer were to remain entirely omniscient, there would be no story to unfold. Thus various points of view are deployed, but the overall effect leans toward "empowering" and flattering the viewer as godlike.

50. Bourdieu, 1984, 152.

51. Claire Johnston, "Towards a Feminist Film Practice." In Nichols, 1985, 323.

52. Tyler Parker, "The Horse: Totem Animal of Male Power." In Gerald Mast and Marshall Cohen, eds. *Film Theory and Criticism* (Oxford: Oxford University Press, 1974), 587.

53. Aumont et al., 1992, 123, 214. Also see Christian Metz, *The Imaginary Signifier: Psychoanalysis and the Cinema* (Bloomington: Indiana University Press 1986), 50, 56, 97. Also see Jean-Louis Baudry, "The Apparatus: Metapsychological Approaches to the Impression of Reality." In Philip Rosen, ed. *Narrative Apparatus, Ideology: A Film Theory Reader* (New York: Columbia University Press, 1986), 313.

54. Sandy Flitterman-Lewis, "Psychoanalysis, Film, and Television." In Robert C. Allen, ed. *Channels of Discourse: Television and Contemporary Criticism* (London: Routledge, 1992), 161.

55. Ibid., 160–63, 171, 178.

56. Pieter J. Fourie, *Media Studies: Content, Audiences, and Production* (Johannesburg: Juta Academic, 2004), 330–31.

57. Aumont et al., 1992, 220; Fourie, 2004, 331.

58. Richard Dyer, *White* (London: Routledge, 1997), passim. Perhaps it is possible, as Edward Said has suggested, that the western literary tradition as well, especially the novel, appeals to—if not actually fosters—middle-class readers' narcissistic tendencies. Said, 1993, 77, 80–96.

59. Tony Yeshin, *Advertising* (Florence, KY: Cengage Learning Business Press, 2005), 3, 378.

60. Yardena Rand, *Wide Open Spaces: Why We Love Westerns* (Manville, RI: Maverick Spirit Press, 2005), 11, 17.

61. Ibid., 13–14; emphasis in original.

62. Richard Abel, "Our Country/Whose Country? The Americanization Project of Early Westerns." In Edward Buscombe and Roberta Peason, eds. *Back in the Saddle Again: New Essays on the Western* (London: BFI Publishing, 1998), 81.

63. Rand (2005) claims western viewers are a "highly diverse" audience, but her own statistics contradict this (12–13, 34). Rand admits to having had misgivings about her own guilty enjoyment of the genre (1, 15), and her book, which began as a doctoral dissertation, works very hard to validate her love of the western despite her being painfully aware that many if not most of her peers perceive the genre as "representing some of the worst tendencies of American culture—an unreconstructed homage to white-male machismo, a zealous exaltation of an imagined history that privileges white men" and "toxic ethnocentrism" (14). This reconciliation presents an unstable and problematic proposition that forms the basis of her study. Certainly it does not offer a monochromatic, simplistic view as do many books celebrating the

western. Rand often cites the fact that many nonwhites, including American Indians themselves, say they enjoy westerns (15, 18, 19, 20, 22, 26, 34, 36). This seems to me another example of the kind of *endorsement* Euro-Americans like to see and hear, endorsement of the very system that has historically oppressed minorities—much like Dr. Veraswami who defends British imperialism in Orwell's *Burmese Days*, 2004, 45–54.

64. Paul Fussell, *Class: A Guide through the American Status System* (New York: Touchstone, 1983), 109.

65. Arthur Shostak, *Blue-Collar Life* (New York: Random House, 1969), 208; emphasis added.

66. In Rand, 2005, 20–21; emphasis added.

67. *CBSNews*. "Why TV Crime Shows Are to Die For." CBSNews.com. May 16, 2010. http://www.cbsnews.com/8301-3445_162-6488704.html.

68. Jason Mittell, *Genre and Television: From Cop Shows to Cartoons in American Culture* (New York: Routledge, 2004), 14; emphasis added.

69. Jack Nachbar and Kevin Lause, "All Together Now: Popular Rituals in America." In Nachbar and Lause, eds. *Popular Culture: An Introductory Text* (Bowling Green, OH: Bowling Green State University Press, 1992), 377–78.

70. Ibid., 381.

71. Bernadette Casey, Neil Casey, Ben Calvert, Liam French, and Justin Lewis, *Television Studies: The Key Concepts* (London: Routledge, 2002), 110.

72. Thomas Schatz, *Hollywood Genres: Formulas, Filmmaking, and the Studio System* (New York: McGraw-Hill, 1981), 47.

73. Casey et al., 2002, 109.

74. Syed M. Khatib (Cedric C. Clark), "The Reduction of Cognitive Conflict through Spatial and Temporal Displacement." *Journal of Intergroup Relations* 15(4), Winter 1987, 28–31.

75. Stephanie Greco Larson, *Media and Minorities: The Politics of Race in News and Entertainment* (Lanham, MD: Rowman & Littlefield, 2006), 52; emphasis added.

76. Thomas Schatz, "The Structural Influence: New Directions in Film-Genre Study." In Barry Keith Grant, ed. *Film Genre Reader IV* (Austin: University of Texas Press, 2003), 116; emphasis added.

77. Media sociologist Todd Gitlin, for example, called the 1977 miniseries *Roots* (based on Alex Haley's best-selling book), which dealt with the effects of slavery and racism, a "national *rite of exorcism* and expiation." Todd Gitlin, *Inside Prime Time* (Berkeley: University of California Press, 2000), 163; emphasis added.

78. George Gerbner, " Television: The New State Religion?" *Et Cetera* 34(2), June 1977, 148.

79. Peter Bogdanovich, *John Ford* (Berkeley: University of California Press, 1978), 104.

80. Hayden White, *Figural Realism: Studies in the Mimesis Effect* (Baltimore: Johns Hopkins University Press, 1999), 69; emphasis added.

81. Mark Golub, "History Died for Your Sins: Guilt and Responsibility in Hollywood Redemption Histories." *Journal of American Culture* 21(3), June 28, 1998, 25.

82. *Soldier Blue*, directed by Ralph Nelson (Katzka-Loeb Productions/Avco-Embassy Pictures/Lionsgate 2006; originally released 1970).

83. Donald L. Kaufman, "The Indian as Media Hand-Me-Down." *Colorado Quarterly* 23, Spring 1975, 497.

84. Jeremy Devine, *Vietnam at 24 Frames a Second: A Critical and Thematic Analysis of over 400 Films about the Vietnam War* (Austin: University of Texas Press, 1999), vvii–xviii, 63, 153, 314. Also see John Ellis, *Seeing Things: Television in the Age of Uncertainty* (London: I. B. Tauris, 2000), 74–129.

85. John Sabini, *Social Psychology* (New York: W. W. Norton, 1992), 626–27.

86. Sigmund Freud, *The Standard Edition of the Complete Psychological Works of Sigmund Freud*. Volume 3. Translated by James Strachey (London: Hogarth Press, 1962), 161–84.

87. Drinnon, 1980, 62–64; Slotkin, 1973, 135–37; Robert F. Berkhofer Jr., *The White Man's Indian: Images of the American Indian from Columbus to the Present* (New York: Alfred A. Knopf, 1978), 83; Peter Hulme, *Colonial Encounters: Europe and the Native Caribbean* (London: Metheun, 1986), 85.

88. David Niose, "Noam Chomsky: On Humanism, the Vulnerability of Secular Nationalism, and the Mother of All Book Plugs." *Humanist* 67(1), January 2007, 25.

89. David Stannard, *American Holocaust: The Conquest of the New World* (Oxford: Oxford University Press, 1992). Also see Ward Churchill, *A Little Matter of Genocide: Holocaust and Denial in the Americas, 1492–Present* (San Francisco: City Lights Books, 1997); also see Lenore A. Stiffarm and Phil Lane Jr., "The Demography of North America: A Question of American Indian Survival." In M. Annette Jaimes, *The State of North America: Genocide, Colonization, and Resistance* (Boston: South End Press, 1992), 23–53.

90. Melbourne Cummings, "The Changing Image of the Black Family on Television," *Journal of Popular Culture* 22, March 5, 2004, 75.

91. Raul Chavez, "Childhood Indians: Television, Film, and Sustaining the White (sub)Conscience." Unpublished PhD dissertation, University of California, Riverside, 2005, 21.

92. James Baldwin, *The Price of the Ticket* (New York: St. Martin's/Marek, 1985), 99.

93. *Images of Indians: How Hollywood Stereotyped the Native American*, directed by Chris O'Brien and Jason Witmer (Starz Encore Entertainment, 2003).

94. Stephanie Larson, 2006, 50.

95. As, for example, the various land-grabs that endured from the 1800s all the way into the 1950s and 1960s. For more specific details of the latter see Richard Drinnon, *Keeper of Concentration Camps: Dillon S. Myer and American Racism* (Berkeley: University of California Press, 1987), 208–13.

96. Ward Churchill, *Struggle for the Land: Native North American Resistance to Genocide, Ecocide, and Colonization* (San Francisco: City Lights Books, 2002), 135–89; also see Greg Palast, "Court Rewards Exxon for Valdez Spill." *Chicago Tribune*, June 26 2008. Reprinted at http://www.gregpalast.com/court-rewards-exxon-for-valdez-oil-spill/?.

97. Annette M. Taylor, "Cultural Heritage in *Northern Exposure*." In S. Elizabeth Bird, ed. *Dressing in Feathers: The Construction of the American Indian in American Popular Culture* (Boulder, CO: Westview Press, 1996), 240; emphasis added.

98. Churchill, 2002, 18.

99. Ralph Friar and Natasha Friar, *The Only Good Indian: The Hollywood Gospel* (New York: Drama Book Specialists, 1972), 2.

100. Todorov, 1984, 42–43.

101. Jean Beaudrillard, *Simulacra and Simulations* (New York: Semiotext[e], 1983), 12–13.

102. Angela Aleiss, *Making the White Man's Indian: Native Americans and Hollywood Movies* (New York: Praeger Publishers, 2005), 62, 69–71, 76.

103. Daniel Chirot and Clark McCauley, *Why Not Kill Them All? The Logic and Prevention of Mass Political Murder* (Princeton, NJ: Princeton University Press, 2010), 54.

104. There are numerous examples throughout history in which individuals or groups of dissidents have sided with invaders against their own countrymen, Jews in Algeria siding with the French being just one example.

105. *Law of the Plainsman*, episode 28, "Jeb's Daughter"; emphasis added. Except for the references to Apaches, this soliloquy could easily have come from the character Gunga Din in the 1939 George Stevens movie of the same name (loosely based on Kipling's 1892 poem).

106. Senator George Malone (R-Nevada) compared Indians to communists in a 1953 U.S. Senate speech in which he calls Indian reservations "natural socialist environments." Senator George Malone, "Emancipation of American Indians." *Congressional Record* 99, July 29, 1953, 10294.

107. Andrew Bacevich, *The Limits of Power* (New York: Metropolitan Books, 2008), 17, 20.

108. There is little question that Protestantism and capitalism are inextricably intertwined. A strong argument could be made that capitalism would not have been possible without the Reformation. See Max Weber, *The Protestant Ethic and the Spirit of Capitalism* (New York: W. W. Norton, 2009), 6.

109. George Brown Tindall, *America: A Narrative History* (New York: W. W. Norton, 1992; originally published 1984), 57.

110. Horsman, 1981, 83. Emphasis added.

111. This was explicitly recognized in Ford's *The Man Who Shot Liberty Valance* (1962).

112. This topic should be explored more thoroughly. Corkin addresses it extensively in *Cowboys as Cold Warriors* (Philadelphia: Temple University Press, 2004).

113. Mordechai Tamarkin, *Cecil Rhodes and the Cape Afrikaaners: The Imperial Colossus and the Colonial Parish Pump* (London: Frank Cass, 1996), 138.

114. What is more, Native American lands were used as dumping grounds for nuclear waste. Randel Hanson, "Indian Burial Grounds for Nuclear Waste." *Multinational Monitor*, September 21, 1995, 1.

Bibliography

Aaronson, James. *The Press and the Cold War* (New York: New York University Press, 1990).

Abel, Richard. "Our Country/Whose Country? The Americanization Project of Early Westerns." In Edward Buscombe and Roberta Peason, eds. *Back in the Saddle Again: New Essays on the Western* (London: BFI Publishing, 1998): 77–95.

"About Sovereign Cherokee Nation Tejas." Texas Cherokee Nation. 2010. http://www.texascherokeenation.org/about.php.

Adorno, Theodor. *The Culture Industry* (London: Routledge, 2006; originally published 1972).

The Adventures of Superman (Motion Pictures for Television, 1952–1958). Arciga, Pepe. *Los Angeles Mirror*, August 26, 1957. Reprinted in *Los Angeles Times*, August 27, 2007. http://latimesblogs.latimes.com/thedailymirror/2007/08/pepe-arciga-1.html#more.

Aleiss, Angela. "Hollywood Addresses Postwar Assimilation: Indian-White Attitudes in *Broken Arrow*." *American Indian Culture and Research Journal* 2(1), October 1987, 609–38.

———. *Making the White Man's Indian: Native Americans and Hollywood Movies* (New York: Praeger Publishers, 2005).

Alexie, Sherman. "I Hated Tonto (Still Do)." *Los Angeles Times*, June 28, 1998. http://articles.latimes.com/1998/jun/28/entertainment/ca-64216.

Althusser, Louis. *Lenin and Philosophy* (London: NLB, 1971; originally published 1968).

"The Angry American Indian: Starting Down the Protest Trail." Time, February 9, 1970. http://www.time.com/time/magazine/article/0,9171,876550,00.html.

Arnold, Elliott. *Blood Brother* (New York: Duell, Sloan & Pearce, 1947).

Arrowhead, directed by Charles Marquis Warren (Paramount Pictures, 1953).

Aufderheide, Pat. "Good Soldiers." In Mark Crispin Miller, ed. *Seeing through Movies: A Pantheon Guide to Popular Culture* (New York: Pantheon Books, 1990), 81–111.

Auld, Kiros. "Pocahontas." Powhatan Museum of Indigenous Arts and Culture. http://www.powhatanmuseum.com/Pocahontas.html.

Aumont, Jacques, Alain Bergala, Michel Marie, and Mark Vernet, eds., *Aesthetics of Film* (Austin: University of Texas Press, 1992).

Avatar, directed by James Cameron (Dune Entertainment/Twentieth Century-Fox Film Corp. 1999).

Bacevich, Andrew. *American Empire: The Realities and Consequences of U.S. Diplomacy* (Cambridge, MA: Harvard University Press, 2002).

———. *The Limits of Power* (New York: Metropolitan Books, 2008).

———. "The Tyranny of Defense Inc." *The Atlantic*, January/February 2011. http://www.theatlantic.com/magazine/archive/2011/01/the-tyranny-of-defense-inc/8342.

Bailey, Courtney W. "Coming Out as Homophobic: Isaiah Washington and the *Grey's Anatomy* Scandal." *Communications and Critical/Cultural Studies* 8(1), March 2011, 1–21.

Baldwin, James. *The Fire Next Time* (New York: Dial Publishing, 1963).

———. *The Price of the Ticket* (New York: St. Martin's/Marek, 1985).

B&H Publishing Group. "Black Belt Patriotism: Chuck Norris." YouTube. http://www.youtube.com/watch?v=kMhLgC8WGz8.

Bannister, Harry. Memo to NBC president Robert Sarnoff. October 17, 1961. Harry R. Bannister collection, State Historical Society of Wisconsin. Reprinted in Mary Ann Watson, *The Expanding Vista: American Television in the Kennedy Years* (Oxford: Duke University Press, 1990).

Barnouw, Erik. *Tube of Plenty: The Evolution of American Television* (Oxford: Oxford University Press, 1990; originally published 1975).

Barsamian, David. "Telling the Truth about Imperialism." *International Socialist Review*, November–December 2003. http://isreview.org/issues/33/humanity.shtml.

Barsh, Russell Lawrence, "Aboriginal Peoples and Canada's Conscience." In Daniel J. K. Beavon, Cora Jane Voyageur, and David R. Newhouse, eds. *Hidden in Plain Sight: Contributions of Aboriginal Peoples to Canadian Identity and Culture*. Volume 1 (Toronto: University of Toronto Press, 2005): 270–92.

Barthel, Joan. "Honest Injun." *New York Times*, July 24, 1966, 77.

Barthes, Roland. *Mythologies* (London: Paladin, 1972; originally published 1957).

———. *The Pleasure of the Text* (New York: Hill and Wang, 1975; originally published 1973), 14.

Barton, Roger. *Media in Advertising* (New York: McGraw-Hill, 1964).

Bataille, Gretchen, ed. *Native American Representations: First Encounters, Distorted Images, and Literary Appropriations* (Lincoln: University of Nebraska Press, 2001).

Bataille, Gretchen M., and Charles L. P. Silet, eds. *The Pretend Indians: Images of Native Americans in the Movies* (Ames: Iowa State University Press, 1980).

Bates, Thomas R. "Gramsci and the Theory of Hegemony." *Journal of the History of Ideas* 36(2), April 1975, 351–66.

Baudry, Jean-Louis. "The Apparatus: Metapsychological Approaches to the Impression of Reality." In Philip Rosen, ed. *Narrative Apparatus, Ideology: A Film Theory Reader* (New York: Columbia University Press, 1986): 299–318.

Bayers, Peter L. "The US Mint, the Lewis and Clark Bicentennial, and the Perpetuation of the Frontier Myth." *Journal of Popular Culture* 44(1), February 2011, 37–52.

Beaudrillard, Jean. *Simulacra and Simulations* (New York: Semiotext[e], 1983).

Bergan, Ronald. "In Praise of Arthur Penn." *Guardian* (United Kingdom), February 14, 2007. http://www.guardian.co.uk/film/filmblog/2007/feb/14/arthurpenn.

Berkhofer, Robert F., Jr. *The White Man's Indian: Images of the American Indian from Columbus to the Present* (New York: Alfred A. Knopf, 1978).

Berkow, Ira. "At Dinner with Chuck Norris." *New York Times*, May 12, 1993, C1.

Bernays, Edward L. *The Engineering of Consent* (Norman: University of Oklahoma Press, 1969).

Bernhard, Nancy. *US Television News and Cold-War Propaganda, 1947–1960* (Cambridge: Cambridge University Press, 2003).

Bernstein, Carl, "The CIA and the Media." *Rolling Stone*, October 20, 1977: 57–61.

The Best of the Lone Ranger: First Three Episodes. Episode 1: "Enter the Lone Ranger" (Apex Films/UAV Corp., 1997; originally aired ABC-TV, September 15, 1949). VHS.

Beveridge, Albert J. Address to U.S. Senate, January 9, 1900. *Congressional Record*, 56th Congress, 1st session, pp. 704–12. Quoted in David Erwin Harrell Jr., Edwin S. Gaustad, and John B. Boles. *Unto a GoodLand: A History of the American People*. Volume 2, *From 1865* (Cambridge: Wm. B. Eerdmans, 2005).

Bhabha, Homi K. "The 'Other' Question: Difference, Discrimination, and the Discourse of Colonialism" in Russell Ferguson, Russell, Martha Gever, Trinh T. Minh-ha, and Cornel West (eds.), *Out There: Marginalization and Contemporary Cultures* (Boston: MIT Press, 1990; originally published in *Screen* 24[6], 1983) 71–87.

Bickel, Mary E. *George W. Trendle: Creator and Producer of the Lone Ranger, the Green Hornet, Sgt. Preston of the Yukon, the American Agent, and Other Successes* (New York: Exposition Press, 1971).

Billy Jack, directed by Tom Laughlin (National Student Film Corp./Warner Bros. Pictures, 1971).

Bird, S. Elizabeth. *Dressing in Feathers: The Construction of the American Indian in American Popular Culture* (Boulder, CO: Westview Press, 1996).

The Birth of a Nation, directed by David W. Griffith (David W. Griffith Corp., 1915).

Biskind, Peter. *Seeing Is Believing: How Hollywood Taught Us to Stop Worrying and Love the Fifties* (New Yok: Henry Holt, 1983).

Bjeljac, David. *American Art: A Cultural History* (New York: Harry N. Abrams, 2001).

Blackburn, Bob L. "From Blood Revenge to the Lighthorsemen: Evolution of Law-Enforcement Institutions among the Five Civilized Tribes in 1861." *American Indian Law Review* 8(1), 1980, 49–63.

Blazing Saddles, directed by Mel Brooks (Crossbow Productions/Warner Bros. Pictures, 1974).

Bly, Robert. *Iron John: A Book about Men* (Boston: Addison-Wesley, 1990).

Bogdanovich, Peter. *John Ford* (Berkeley: University of California Press, 1978).

Bourdieu, Pierre. *Distinction: A Social Critique of the Judgment of Taste* (Cambridge, MA: Harvard University Press, 1984; originally published 1979), 152, 214, 598.

Borstelmann, Thomas. *The Cold War and the Color Line: American Race Relations in the Global Arena* (Cambridge, MA: Harvard University Press, 2001).

Boyle, Brenda M. "Rescuing Masculinity." *Journal of American Culture* 34(2), June 2011, 149–60.

Brave Eagle (Frontier Productions/CBS Television, 1955). DVD copy from unknown broadcast.

"Brave Eagle, 1955." Turner Classic Movies. http://turnerclassic.moviesunlimited.com/product.asp?sku=D28770.

Brave Warrior, directed by Spencer Gordon Bennett (Esskay Pictures, 1952).

Broken Arrow (series) (Twentieth Century-Fox Pictures/ABC-television, 1956–1958).

Broken Arrow, directed by Delmer Daves (Twentieth Century-Fox Pictures, 1950). DVD copy from unknown source.

Brooks, Tim, and Earl Marsh. *The Complete Directory to Prime Time Network and Cable Television Shows* (New York: Random House, 2003).

Brown, Oscar, Jr. "The Lone Ranger." *Fresh* (Atlantic Records, SD 18106, 1974). Edward Buscombe, and Roberta Peason, eds. *Back in the Saddle Again: New Essays on the Western* (London: BFI Publishing, 1998).

Bryant, John, ed. *American Literature* (Boston: Pearson Custom Publications, 2003).

"Buffy Sainte-Marie" (biography). CreativeNative.com. 2003. http://www.creative-native.com/biograp.htm.

Burt, Ramsay. *The Male Dancer: Bodies, Spectacle and Sexuality* (London: Routledge, 1995).

"Burt Reynolds Should Have Played John Hawk More Like Quint Asper and Less Like Ben Casey." *Reviews and Ratings for "Hawk"* (1966). Internet Movie Database, June 11, 2006. http://www.imdb.com/title/tt0059992/reviews.

Buscombe, Edward. *100 Westerns: BFI Screen Guides* (London: British Film Institute, 2006).

Buxton, David. *From "The Avengers" to "Miami Vice": Form and Ideology in Television* (Manchester: Manchester University Press, 1990).

Cade's County (David Gerber Productions/Twentieth Century-Fox Television/CBS-TV, 1971).

Campbell, Christopher, and Lori Edmo Suppah. "Television's Troubling Indian Images." *Television Quarterly* 33(4), Spring 2003, 16–26.

Capstone Press Geography Department. *Arizona: One Nation* (Mankato, MN: Capstone Press, 2002).

"Carlos Ray Norris a.k.a. Chuck Norris Family History." RootsWeb. http://wc.rootsweb.ancestry.com/cgi-bin/igm.cgi?op=GET&db=14x24&id=I0141&style=TABLE.

Carmichael, Deborah A. *The Landscape of Hollywood Westerns: Eco-criticism in an American Film Genre* (Salt Lake City: University of Utah Press, 2006).

Carrier, Lyman. *The Beginnings of Agricultre in America* (New York: McGraw-Hill, 1923).

Carstarphen, Meta G., and John P. Sanchez, eds. *American Indians in the Mass Media* (Norman: University of Oklahoma Press, 2012).

Casey, Bernadette, Neil Casey, Ben Calvert, Liam French, and Justin Lewis. *Television Studies: The Key Concepts* (London: Routledge, 2002).

CBSNews. "Why TV Crime Shows Are to Die For." CBSNews.com. May 16, 2010. http://www.cbsnews.com/8301-3445_162-6488704.html.

Champagne, Duane. *Contemporary Native American Cultural Issues* (Lanham, MD: Rowman & Littlefield, 1999).

Chapman, James, and Nicholas J. Cull. *Projecting Empire: Imperialism and Popular Cinema* (London: I. B. Tauris, 2009).

Chapman, Roger. *Culture Wars: An Encyclopedia of Issues, Viewpoints, and Voices*. Volume 1 (Armonk, NY: M. E. Sharpe, 2010).

Chavez, Raul. "Childhood Indians: Television, Film, and Sustaining the White (sub)Conscience." Unpublished PhD dissertation, University of California, Riverside, 2005.

Chen, Chiung Hwang. *Mormon and Asian-American Model-Minority Discourses in News and Popular Magazines* (Lewiston, NY: Edwin Mellen Press, 2004).

Cherry, Conrad. *God's New Israel: Religious Interpretations of American Destiny* (Englewood Cliffs, NJ: Prentice-Hall, 1971).

Cheyenne (Warner Bros. Television/ABC-TV, 1955).

Cheyenne Autumn, directed by John Ford (Ford-Smith Productions/Warner Bros. Pictures, 1964).

Chirot, Daniel, and Clark McCauley. *Why Not Kill Them All? The Logic and Prevention of Mass Political Murder* (Princeton, NJ: Princeton University Press, 2010).

Chomsky, Noam. *Necessary Illusions: Thought Control in Democratic Societies* (Boston: South End Press, 1989).

Churchill, Ward. *Fantasies of the Master Race: Literature, Cinema, and the Colonization of American Indians* (San Francisco: City Lights Books, 1998; originally published 1992).

———. *A Little Matter of Genocide: Holocaust and Denial in the Americas, 1492–Present* (San Francisco: City Lights Books, 1997).

———. *Struggle for the Land: Native North American Resistance to Genocide, Ecocide, and Colonization* (San Francisco: City Lights Books, 2002).

Churchill, Ward, Mary Anne Hill, and Norbert S. Hill Jr. "Examination of Stereotyping: An Analytical Survey of Twentieth Century Indian Entertainers." In Gretchen M. Bataille and Charles L. P. Silet, eds. *The Pretend Indians: Images of Native Americans in the Movies* (Ames: Iowa State University Press, 1980): 35–48.

Cirino, Robert. *We're Being More Than Entertained* (Honolulu: Lighthouse Press, 1977).

Clark, Cedric C. "The Concept of Legitimacy in Black Psychology." In Edgar G. Epps, ed. *Race Relations: Current Perspectives* (Cambridge, MA: Winthrop, 1973): 332–54.

——— (Syed M. Khatib). "Television and Social Control: Some Observations on the Portrayals of Ethnic Minorities." *Television Quarterly*, Spring 1969, 18–22.

Cooper, James Fenimore. *The Leatherstocking Tales*. Volume 1 (New York: Library of America, 1985; originally published 1827–1841).

Corkin, Stanley. *Cowboys as Cold Warriors* (Philadelphia: Temple University Press, 2004).

Cornell, Drucilla. *Clint Eastwood and Issues of American Masculinity* (New York: Fordham University Press, 2009).

Cosby, Bill. "The Lone Ranger." *I Started Out as a Child* (Warner Bros. Records, WS 1567, 1964).

Cosgrove, Vincent. "A Tubful of Coonskin and Corn." *New York Times*, October 29, 2006. http://www.nytimes.com/2006/10/29/arts/television/29cosg.html?_r=0.

Cowles, C. S., Stanley N. Gundry, Daniel L. Gard, Eugene H. Merrill, and Tremper Longman. *Show Them No Mercy: Four Views of Canaanite Genocide* (Grand Rapids, MI: Zondervan, 2003).

Cox, Mike. "A Brief History of the Texas Rangers." Texas Ranger Hall of Fame and Museum. http://www.texasranger.org/history/BriefHistory1.htm.

Cripps, Thomas. *Making Movies Black: The Hollywood Message Movie from World War II to the Civil Rights Era* (Oxford: Oxford University Press, 1993).

Crofts, Stephen. "Authorship and Hollywood." In John Hill and Pamela Church Gibson, eds. *The Oxford Guide to Film Studies* (Oxford: Oxford University Press, 1998), 310–26.

Crook, George. *Resumé of Operations against Apache Indians, 1882 to 1886* (London: Johnson-Taunton Military Press, 1971; originally published 1886).

Cummings, Melbourne. "The Changing Image of the Black Family on Televison." *Journal of Popular Culture* 22, March 5, 2004, 75–85.

Custalow, Linwood, and Angela L. Daniel. *The True Story of Pocahontas* (Golden, CO: Fulcrum Publishing, 2007).

Custen, George F. *Twentieth Century's Fox: Darryl F. Zanuck and the Culture of Hollywood* (New York: Basic Books, 1997).

Dances with Wolves, directed by Kevin Costner (Tig Productions/Majestic Films International/Orion Pictures, 1990).

Daniel Boone (Arcola Pictures/Fespar Productions/Twentieth Century-Fox Pictures/NBC-TV, 1964–1970).

Daniel Boone: The Best of Mingo (Arcola Pictures/Fespar Productions/Twentieth Century-Fox Pictures/Liberation Entertainment, 2009; originally aired 1964–1970). DVD.

Davy Crockett, King of the Wild Frontier (Walt Disney Productions/ABC-TV, 1955).

Defoe, Daniel. *Robinson Crusoe* (London: CRW Publishing, 2007; originally published 1719).

DeLong, Anne. "Disrupting the Discourse of Conquest: The Suppression of Sepúlveda." The Literature of Justification. Accessed August 1, 2011. Lehigh University Digital Library. http://digital.lib.lehigh.edu/trial/justification/newspain/essay/.

Deloria, Philip J. *Playing Indian* (New Haven, CT: Yale University Press, 1998).

Deloria, Vine, Jr. *Custer Died for Your Sins: An Indian Manifesto* (Norman: University of Oklahoma Press, 1988; originally published 1969).

———. *We Talk You Listen: New Tribes, New Turf* (Lincoln, NE: Bison Books, 2007; originally published 1970).

Deloria, Vine, Jr. in Gretchen M. Bataille and Charles L. P. Silet (eds.), *The Pretend Indians: Images of Native Americans in the Movies* (Ames: Iowa State University Press, 1980), 189.

Delta Force, directed by Menahem Golan (Cannon Films/MGM Pictures, 1986).

der Derian, James. *Virtuous War: Mapping the Military-Industrial-Media-Entertainment Complex* (New York: Routledge, 2009; originally published 2001).

Devine, Jeremy. *Vietnam at 24 Frames a Second: A Critical and Thematic Analysis of over 400 Films about the Vietnam War* (Austin: University of Texas Press, 1999; originally published 1995).

Dorfman, Ariel. *The Emperor's Old Clothes: What the Lone Ranger, Babar, and Other Heroes Do to Our Minds* (Durham, NC: Duke University Press, 2010; originally published 1983).

Dowd, Maureen. "Cuomos vs. Sopranos." *New York Times,* April 22, 2001, WK17.

Dower, John W. *War without Mercy: Race and Power in the Pacific War* (New York: Pantheon, 1986).

Dragnet (Mark VII Productions/NBC-TV, 1951–1959; 1967–1970).

Drinnon, Richard. *Facing West: The Metaphysics of Indian-Hating and Empire-Building* (Minneapolis: University of Minnesota Press, 1980).

———. *Keeper of Concentration Camps: Dillon S. Myer and American Racism* (Berkeley: University of California Press, 1987).

Dr. Quinn, Medicine Woman (Paramount Pictures/CBS-TV, 1993).

DuBois, W. E. B. *Darkwater: Voices from within the Veil. The Oxford W.E.B. DuBois Reader*. Edited by E. J. Sundquist (Oxford: Oxford University Press, 1996; originally published 1920).

———. *The Souls of Black Folk* (Rockville, MD: Arc Manor, 2008; originally published 1903).

Dudziak, Mary L. *Cold War Civil Rights: Race and the Image of American Democracy* (Princeton, NJ: Princeton University Press, 2000).

Durham, Jimmy, "Cowboys and . . ." In M. Annette Jaimes, ed. *The State of Native America: Genocide, Colonization, and Resistance* (Boston: South End Press), 423–38.

Dyer, Richard. "The Role of Stereotypes." In Paul Marris and Sue Thornham, eds. *Media Studies: A Reader* (New York: New York University Press, 2000; originally published 1996): 245–51.

———. *White* (London: Routledge, 1997).

Eakin, Emily. "All Roads Lead to D.C." *New York Times*, Week in Review, March 31, 2002. http://www.nytimes.com/learning/teachers/featured_articles/20020401monday.html.

Eckstein, Arthur M., and Peter Lehman. *The Searchers: Essays and Reflections on John Ford's Classic Western* (Detroit: Wayne State University Press, 2004).

Edwards, Paul M. *The Korean War: A Historical Dictionary* (Lanham, MD: Scarecrow Press, 2010).

Eldridge, David. "'Dear Owen': The CIA, Luigi Luraschi and Hollywood, 1953." *Historical Journal of Film, Radio, and Television* 20(2), June 2000, 51–156.

Ellis, John. *Seeing Things: Television in the Age of Uncertainty* (London: I. B. Tauris, 2000).

Ely, Melvin P. *The Adventures of Amos 'n' Andy: A Social History of an American Phenomenon* (New York: Free Press, 1991).

Erdoes, Richard, and Alfonso Ortiz. *American Indian Myths and Legends* (New York: Random House, 1985).

Erickson, Hal. "Michael Ansara: Biography." All Movie Guide. Accessed February 28, 2009. http://www.allmovie.com/cg/avg.dll.

"Famous Masons." Scottish Rite Masonic Center, Valley of San Diego. May 2009. http://www.scottishritesandiego.org/famousmasons.html.

Fanon, Frantz. *Black Skin, White Mask* (New York: Grove Press, 2008; originally published 1952).

———. *The Wretched of the Earth* (New York: Grove Press, 2004; originally published 1961).

Faragher, John Mack. *The Life and Legend of an American Pioneer* (New York: Henry Holt, 1993).

Farrell, Kirby. "The Berserk Style in American Culture." *Cultural Critique* 46, Autumn 2000, 179–209.

Fausz, J. Frederick. "Opechancanough: Indian Resistance Leader." In Gary Nash and David Sweet, eds. *Struggle and Survival in Colonial America* (Berkeley: University of California Press, 1981): 21–37.

Feiffer, Jules. "The Minsk Theory of Krypton." *New York Times Magazine*, December 29, 1996, 14–15.

Festinger, Leon. *A Theory of Cognitive Dissonance* (Stanford, CA: Stanford University Press, 1957).

Fiedler, Leslie. *The Return of the Vanishing American* (New York: Stein & Day, 1968).

Firewalker, directed by J. Lee Thompson (Cannon Films, 1986).

First Blood, directed by Ted Kotcheff (Anabasis N.V./Orion Pictures, 1982).

Fiske, John. *Television Culture* (London: Routledge 1989; originally published 1987).

FitzGerald, Michael Ray. "Evolutionary Stages of Minorities in the Mass Media: An Application of Clark's Model to American Indian Representations." *Howard Journal of Communications* 21(4), November 2010, 367–84.

———. "Sitcoms and Suburbia: The Role of Network Television in the De-urbanization of the US." Unpublished master's thesis, University of Florida, 2007.

Flitterman-Lewis, Sandy. "Psychoanalysis, Film, and Television." In Robert C. Allen, ed. *Channels of Discourse: Television and Contemporary Criticism* (London: Routledge, 1992), 203–46.

Ford, Glen. "Corporate Media's Crimes against Libya—and Humanity." *Black Agenda Report*, June 8, 2011. http://www.trinicenter.com/modules.php?name=AvantGo&op= ReadStory&sid=2332.

Forte, Maximilian C., ed. *The New Imperialism: Militarism, Humanism, and Occupation.* Montreal: Alert Press, 2010).

Foster, James H. *The Rungless Ladder: Harriet Beecher Stowe and New England Puritanism* (New York: Cooper Square Publishers, 1970).

Foster, John Bellamy, Hannah Holleman, and Robert W. McChesney, "The US Imperial Triangle and Military Spending." *Monthly Review* 60(5), October 2008. http://monthlyreview.org/081001foster-holleman-mcchesney.php.

Fourie, Pieter J. *Media Studies: Content, Audiences, and Production* (Johannesburg: Juta Academic, 2004).

Fowke, Gerard. *Archeological History of Ohio: The Mound Builders and Later Indians* (Columbus: Ohio State Archaeological and Historical Society, 1902).

Francis, Daniel. "The Imaginary Indian." In Tania Das Gupta, Carl E. James, Roger C. A. Maaka, Grace Edward Galabuzi, and Chris Anderson, eds. *Race and Racialization: Essential Readings* (Toronto: Canadian Scholars Press, 2007), 234–39.

———. *The Imaginary Indian: The Image of the Indian in Canadian Culture* (Vancouver, BC: Arsenal Pulp Press, 1992).

Franits, Wayne. *Dutch Seventeenth-Century Genre Painting: Its Stylistic and Thematic Evolution* (New Haven, CT: Yale University Press, 2004).

French, Philip. "The Indian in the Western Movie." In Gretchen M. Bataille and Charles L. P. Silet, eds. *The Pretend Indians: Images of Native Americans in the Movies* (Ames: Iowa State University Press, 1980): 98–105.

———. *Westerns* (New York: Viking Press, 1973).

Freud, Sigmund. *The Standard Edition of the Complete Psychological Works of Sigmund Freud.* Volume 3. Translated by James Strachey (London: Hogarth Press, 1962; originally published 1912).

Friar, Ralph, and Natasha Friar. *The Only Good Indian: The Hollywood Gospel* (New York: Drama Book Specialists, 1972).

Friendly, Fred. *The Good Guys, the Bad Guys, and the First Amendment: Free Speech vs. Fairness in Broadcasting* (New York: Random House, 1975).

F-Troop (Warner Bros. Television/ABC-TV, 1965–1967).

Fulbright, William J. Remarks given on the Senate floor. U.S. Congress Senate. Committee on Foreign Relations. August 1966. Reprinted at Center for Research Libraries. http://images.crl.edu/178.pdf.

"Full Cast and Crew for *Broken Arrow*." Internet Movie Database. Accessed April 2, 2009. http://www.imdb.com/title/tt0048848/fullcredits#cast.

Fussell, Paul. *Class: A Guide through the American Status System* (New York: Touchstone, 1983).

Garland, Phyl. "Blacks Challenge the Airwaves." *Ebony* 21(1), November 1970, 35–44.

Garrison's Gorillas (Selmur Productions/ABC-TV, 1967–1968).

Gates, David. "White Male Paranoia." *Newsweek*, March 29, 1993, 48–53.

George, Diana, and Susan Sanders. "Reconstructing Tonto: Cultural Formations and American Indians in 1990s Television Fiction." *Cultural Studies* 9(3), 1995, 427–52.

Gerbner, George. *Casting and Fate: Women and Minorities on Television Drama, Game Shows, and News*. Screen Actors Guild report, 1993. http://www.asc.upenn.edu/gerbner/Asset.aspx?assetID=1074.

———. " Television: The New State Religion?" *Et Cetera* 34(2), June 1977, 192–95.

———. *Women and Minorities on Television: A Study in Casting and Fate*. Screen Actors Guild report, 1993. http://www.asc.upenn.edu/gerbner/Asset.aspx?assetID=333.

Gerbner, George, and Larry Gross. "Living with Television: The Violence Profile. *Journal of Communication* 26(2), June 1976, 172–94.

Ghost of Zorro, directed by Fred C. Brannon (Republic Pictures, 1949).

Gilmore, Richard."The Dark Sublimity of *Chinatown*." In Mark T. Conard, ed. *The Philosophy of Neo-Noir* (Lexington: University Press of Kentucky, 2007), 119–36.

Gitlin, Todd. *Inside Prime Time* (Berkeley: University of California Press, 2000; originally published 1983).

———. "Prime Time Ideology: The Hegemonic Process in Television Entertainment." In Horace Newcomb, ed. *Television: The Critical View* (Oxford: Oxford University Press, 2007), 574–94.

———. *Watching Television: A Pantheon Guide to Popular Culture* (New York: Pantheon, 1986).

Goff, Stan. "An Open Letter to GIs: Hold On to Your Humanity." *International Socialist Review*, January 2004, http://isreview.org/issues/33/humanity.shtml.

Golub, Mark. "History Died for Your Sins: Guilt and Responsibility in Hollywood Redemption Histories." *Journal of American Culture*, 21(3), June 28, 1998, 23–45.

Gould, Jack. "TV: NBC Tarzan, He Urbane and Sophisticated." *New York Times*, September 9, 1966, 91. ProQuest Historical Newspapers.

"Govt. Gives Plans of B'caster Role in Civil Defense." *Variety*, April 18, 1951, 4.

Gramsci, Antonio. *Selections from the Prison Notebooks* (New York: International Publishers, 1971; written 1929).

Grant, Barry Keith, ed. *Film Genre Reader IV* (Austin: University of Texas Press, 2003; originally published 1986).

Gray, Herman. "The Politics of Representation in Network Television." In Horace Newcomb, ed. *Television: The Critical View*, 6th ed. (Oxford: Oxford University Press, 2000), 294–95.

Greider, William. "Acceptable Villains Make Our Troubles So Manageable." *Washington Post*, July 15, 1979, E1, 4.

Grossberg, Larry. "Cultural Studies vs. Political Economy: Is Anybody Else Bored with This Debate?" In John Storey, ed. *Cultural Theory and Popular Culture: A Reader* (Athens: University of Georgia Press, 1998; originally published 1994): 613–24.

Guerrero, Ed. "The Black Man on Our Screens and the Empty Space in Representation." *Callaloo* 18(2), 1995, 395–400.

Gunsmoke (Arness Production Co./CBS-Television, 1955–1975).

Gupta, Tania Das, Carl E. James, Roger C. A. Maaka, Grace Edward Galabuzi, and Chris Anderson, eds. *Race and Racialization: Essential Readings* (Toronto: Canadian Scholars Press, 2007).

Hage, Ghassan. *White Nation: Fantasies of White Supremacy in a Multicultural Society* (Annandale, NSW: Pluto Press, 1998).

Hall, Stuart. "Notes on Deconstructing the Popular." In Raphael Samuel, ed. *People's History and Socialist Theory* (London: Routledge, 1981), 227–40.

———. "The Question of Cultural Identity." In Stuart Hall, David Held, and Tony McGrew, eds. *Modernity and Its Future* (Cambridge: Polity Press, 1992), 274–316.

Hallin, Daniel C. "We Keep America on Top of the World." In Todd Gitlin, ed. *Watching Television: A Pantheon Guide to Popular Culture* (New York: Pantheon, 1986): 9–41.

Hannity & Colmes (Fox News Network, June 20, 2001).

Hanson, Randel. "Indian Burial Grounds for Nuclear Waste." *Multinational Monitor*, September 21, 1995, 1.

Harrell, David Edwin, Jr., Edwin S. Gaustad, and John B. Boles. *Unto a Good Land: A History of the American People*. Volume 2, *From 1865* (Cambridge: Wm. B. Eerdmans, 2005).

Hasse, Larry J. "Termination and Assimilation: Federal Indian Policy, 1943–1961." Unpublished PhD thesis, Washington State University, 1974.

Hawk (Screen Gems/ABC-Television, 1966).

Hayward, Susan. *Cinema Studies: The Key Concepts* (London: Routledge, 2000).

Hedges, Chris. *American Fascists: The Christian Right and the War on America* (New York: Free Press, 2006).

Herman, Edward S., and Noam Chomsky. *Manufacturing Consent: The Political Economy of the Mass Media* (New York: Pantheon Books, 1988).

Hilger, Michael. *The American Indian in Film* (Metuchen, NJ: Scarecrow Press, 1986).

Hill, John, and Pamela Church Gibson, eds. *The Oxford Guide to Film Studies* (Oxford: Oxford University Press, 1998).

Hillerman, Tony. *The Blessing Way* (New York: HarperTorch, 1990; originally published 1970).

"History of the Breed." American Paint Horse Association. Accessed May 1, 2008. http://www.apha.com/breed/history.html.

Hobson, Geary. *The Remembered Earth: An Anthology of Native American Literature* (Albuquerque: University of New Mexico Press, 1981).

Hodges, Ann. "Money Woes May Sideline Norris Series." *Houston Chronicle*, April 14, 1993, 1.

Hoffman, Donald. "Whose Home on the Range? Finding Room for Native Americans, African Americans, and Latino Americans in the Revisionist Western." *MELUS* 22(2), Summer 1997, 45–60.

Holm, Tom. "Patriots and Pawns: State Use of American Indians in the Military and the Process of Nativization of the United States." In M. Annette Jaimes, ed. *The State of Native America: Genocide, Colonization, and Resistance* (Boston: South End Press, 1992): 345–70.

Holst, Jerome A. TVAcres. http://tvacres.com.

Homans, Peter. "Puritanism Revisited: An Analysis of the Contemporary Screen-Image Western" In Jack Nachbar, ed. *Focus on the Western* (Englewood Cliffs, NJ: Prentice-Hall, 1974): 84–92.

Hondo (Andrew J. Fenady Productions/Batjac Productions/MGM Television/ABC-TV, 1967).

Hondo, directed by John Farrow (Wayne-Fellows Productions/Paramount Pictures, 1953).

"*Hondo*: Episode Guide." TV.com. http://www.tv.com/Hondo/show/5143/episode.html.

Horsman, Reginald. *Race and Imperialism* (Cambridge, MA: Harvard University Press, 1981).

Hoy, Suellen. *Chasing Dirt: The American Pursuit of Cleanliness* (Oxford: Oxford University Press, 1995).

Hughes, Mike. "Chuck Norris Finds Television and Fatherhood Better Second Time Around." *Tulsa World*, October 16, 2005, TV5. ProQuest Newspapers.

Hulme, Peter. *Colonial Encounters: Europe and the Native Caribbean* (London: Metheun, 1986).

Hyde, Lewis. *The Gift: Imagination and the Erotic Life of Property* (New York: Vintage, 2007).

Images of Indians: How Hollywood Stereotyped the Native American, directed by Chris O'Brien and Jason Witmer (Starz Encore Entertainment, 2003).

The Incredible Hulk (Universal Television/CBS-TV, 1978–1982).

In the Heat of the Night (series). Fred Silverman Productions/JADDA Productions/Juanita Bartlett Production/NBC-TV, 1988–1994.

In the Heat of the Night, directed by Norman Jewison (Mirisch Corp./United Artists, 1967).

Jackson, Wayne. *Select Studies in the Book of Revelation* (Stockton, CA: Courier Publications, 1995).

Jackson Committee. "March Meeting Notes." April 10, 1953. Washburn Abbott Papers, Box 7. Dwight D. Eisenhower Library, Abilene, Kansas.

Jacobs, Randy D. "Advertising Agencies and the Adoption of Radio: A Diffusion of Innovations Perspective." Paper presented at Conference on Historical Analyses and Research in Marketing, New York, May 22, 2011.

Jenkins, Tricia. *The CIA and Hollywood: How the Agency Shapes Film and Television* (Austin: University of Texas Press, 2012).

———. "Get Smart: A Look at the Current Relationship between Hollywood and the CIA." *Historical Journal of Film, Radio and Television* 29(2), June 2009, 229–43.

———. "Nationalism and Gender: The 1970s, *The Six Million Dollar Man*, and *The Bionic Woman*." *Journal of Popular Culture* 44(1), February 2011, 93–113.

Jennings, Francis. *The Invasion of America: Indians, Colonialism and the Cant of Conquest* (Chapel Hill: University of North Carolina Press, 1975).

Jewett, Robert, and John Shelton Lawrence. *The American Monomyth* (Lanham, MD: University Press of America, 1988).

Johansen, Bruce E. *Forgotten Founders: How the American Indian Helped Shape Democracy* (Cambridge, MA: Harvard Commons Press, 1987).

Johnson, Dirk. "Spiritual Seekers Borrow Indians' Ways." *New York Times*, December 27, 1993, A1.

Johnston, Claire. "Towards a Feminist Film Practice." In Bill Nichols, ed. *Movies and Methods*. Volume 2 (Berkeley: University of California Press, 1985), 323.

Jones, Jeff. *How to Photograph and Edit Women* (Raleigh, NC: Lulu.com, 2006), 71.

Jones, K. Randel. *In the Footsteps of Daniel Boone* (Winston-Salem: John F. Blair, 2005).

"KAB Public-Service Announcements." Keep America Beautiful. https://secure2.convio.net/kab/site/SPageServer?pagename=kabmedia_psaandJServSessionIdr006=iqn2rv6652.app6b.

Kane, Bob. "Legend: The Bat Man and How He Came to Be." *Detective Comics* 33, November 1939. Excerpted in "Detective Comics No. 33, November 1939: 'The Bat Man Wars against the Dirigible of Doom.'" Detective Comments: Batman in Detective Comics. http://www.detectivecomments.com/chronology/1939-11.html.

Kaplan, Richard L. "Spider Man in Love: A Psychoanalytic Interpretation." *Journal of Popular Culture* 44(2), April 2011, 291–313.

Kaufman, Donald L. "The Indian as Media Hand-Me-Down." *Colorado Quarterly* 23, Spring 1975, 489–504.

Kay, Jeanne. "Human Dominion over Nature in the Hebrew Bible." *Annals of the Association of American Geographers* 79(2), June 1989, 214–32.

Kelly, Robin D. G. "A Poetics of Anticolonialism." *Monthly Review* 5(6), November 1999. http://www.monthlyreview.org/1199kell.htm.

Kendrick, James. *Fighting Outward, Looking Inward* (Carbondale: Southern Illinois University Press, 2009).

Kerkeslager, Allen. "Apollo, Greco-Roman Prophecy, and the Rider on the White Horse in Revelation 6:2." *Journal of Biblical Literature* 112(1), Spring 1993, 116–21.

Kerner Commission. *Report of the National Advisory Commission on Civil Disorders* (Washington: U.S. Government Printing Office, 1968).

Khatib, Syed M. (Cedric C. Clark). "The Reduction of Cognitive Conflict through Spatial and Temporal Displacement." *Journal of Intergroup Relations* 15(4), Winter 1987, 28–31.

King, Susan. "In Secret Service of Fun." *Los Angeles Times*, November 4, 2008. http://articles.latimes.com/ 2008/nov/04/entertainment/et-wildwildwest4.

King of the Hill (Deedle-Dee Productions/Fox Television, 1997–2008).

Kipling, Rudyard. "(Take Up) The White Man's Burden." *McClure's* 12, February 12, 1899. Reprinted at World Civilizations Reader, http://www.wsu.edu:8080/~wldciv/world_civ_reader/world_civ_reader_2/kipling.html.

Knapp, Bettina L. *Women, Myth, and the Feminine Principle* (Albany: State University of New York Press, 1997).

Koepnick, Lutz P. "Siegfried Rides Again: Westerns, Technology, and the Third Reich." *Cultural Studies* 11(3), 1997, 418–42.

Kraszewski, Jon. "Adapting Scripts in the 1950s: The Economic and Political Incentives for Television Anthology Writers." *Journal of Film and Video* 58(3), Fall 2006, 3–22.

"Landscape Painting." National Gallery of Art. 2011. http://www.nga.gov/education/american/landscape.shtm.

Larson, Charles. "'The Non-Person,' Final Draft" (episode script). *Nakia* (David Gerber Productions, June 13, 1974).

Larson, Stephanie Greco. *Media and Minorities: The Politics of Race in News and Entertainment* (Lanham, MD: Rowman & Littlefield, 2006).

Lasch, Christopher. *The Culture of Narcissism: American Life in an Age of Diminishing Expectations* (New York: W. W. Norton, 1979).

Lasn, Kalle. *Culture Jam: The Uncooling of America* (New York: William Morrow, 1999).

Lasswell, Harold D. "Information Policies Concerning Russia (Rough Draft)." Memo to William Benton. October 22, 1945. National Archives RG59, Assistant Secretary for Public Affairs, Box 13, Correspondence, 1945–1948.

Last of the Mohicans: Director's Expanded Edition, directed by Michael Mann (Morgan Creek Productions/Twentieth Century-Fox Films, 1999; originally released 1992). DVD.

Laurent, Laurence. "Law 'n' Order Still Dominates New Schedule." *Washington Post*, July 28, 1974, TV-7.

Law and Order: Special Victims Unit (Wolf Films/NBC-Universal Television/Studios USA Television, 1999–).

Law of the Plainsman (Cardiff Productions/Four Star Television/NBC-TV, October 15, 1959). Rerun aired on Lonestar Network, Canada. DVD copy.

Lawrence, John Shelton, and Robert Jewett. "Interplay between Popular Myth and American Foreign Policy." Paper presented at the American Culture Association, New Orleans, LA, April 17, 2003.

Lawrence of Arabia, directed by David Lean (Horizon Pictures/Columbia Pictures, 1963).

Lee, Eloise, and Robert Johnson. "The 25 Biggest Defense Companies in America." *Business Insider*, March 13, 2012. http://www.businessinsider.com/top-25-us-defense-companies-2012-2?op=1.

Lee, Stacey J. *Unraveling the Model-Minority Stereotype* (New York: Teacher's College Press, 1996).

Lenihan, John H. *Showdown: Confronting Modern America in the Western Film* (Champaign: University of Illinois Press, 1980).

Lens, Stanley. *The Forging of the American Empire: From the Revolution to Vietnam, a History of U.S. Imperialism* (London: Pluto Press, 2003; originally published 1971).

Leviero, Anthony. "President Declares a National Emergency." *New York Times*, December 17, 1950, A1. ProQuest Historical Newspapers.

Levine, Richard M. "Counting Heads: TV Demographics." *APF Reporter* 1(3). http://www.aliciapatterson.org/APF0103/Levine/Levine.html.

Lewin, Tamar. "Report Takes Aim at 'Model Minority' Stereotype of Asian-American Students." *New York Times*, June 10, 2008. http://www.nytimes.com/2008/06/10/education/10asians.html.

Lewis, David Rich. "Still Native: The Significance of Native Americans in the History of the Twentieth Century American West." *Western Historical Quarterly* 2(24), May 1993, 203–28.

Lewis, Orrin. "Setting the Record Straight About Native Peoples." NativeLanguages.org. http://www.native-languages.org/iaq10.htm.

Lewis, Richard Warren. "Candid Conversation: John Wayne." *Playboy* 209, May 1971, 78. Reprinted in Barry G. Golson, ed. *The Playboy Interview* (New York: Playboy Press, 1981): 203–27.

Lewis, R. W. B. *The American Adam: Innocence, Tragedy, and Tradition in the Nineteenth Century* (Chicago: University of Chicago Press, 1955).

Linder, Doug. "Lenny Bruce." University of Missouri, Kansas City, Faculty Projects. 2003. http://law2.umkc.edu/faculty/projects/ftrials/bruce/brucebiography.html.

Lippmann, Walter. *Public Opinion* (New Brunswick, NJ: Transaction Publishers, 1998).

Little Big Man, directed by Arthur Penn (Cinema Center Films/Paramount Pictures, 1970).

"Living Conditions." Native American Aid. http://www.nrcprograms.org/site/PageServer?pagename=naa_livingconditions.

Loehr, Davidson. *America, Fascism, and God: Sermons from a Heretical Preacher* (White River Junction, VT: Chelsea Green, 2005).

The Lone Ranger, directed by Stuart Heisler (Lone Ranger Pictures/VCI Entertainment, 2002; originally released by Warner Bros., 1956).

The Lone Ranger. Episode 1: "Enter the Lone Ranger" (Apex Film Corp./ABC-TV, September 15, 1949).

The Lone Ranger and the Lost City of Gold, directed by Lesley Selander (Lone Ranger Pictures/VCI Entertainment, 2002; originally released by United Artists 1958). DVD.

Lone Wolf McQuade, directed by Steve Carver (Orion Pictures, 1983).

Longmire (Warner Horizon Television/A&E, 2012).

Lovett, Lyle. "If I Had a Boat." *Pontiac* (Curb Records, 471795, 1987).

Ludescher, S. J. "Report: Navajos lag behind US." *Gallup Independent* (New Mexico), July 19, 2009. http://www.gallupindependent.com/1999-2001/7-18-00.html.

MacDonald, J. Fred. *Television and the Red Menace: The Video Road to Vietnam* (Westport, CT: Praeger, 1985).

———. *Who Shot the Sheriff? The Rise and Fall of the Television Western* (Westport, CT: Praeger, 1987; originally published 1986).

MacKee, Marty. "Plot Summary for *Hawk* (1966)." Internet Movie Database. http://www.imdb.com/title/tt0059992/plotsummary.

Malone, George. "Emancipation of American Indians." *Congressional Record* 99, July 29, 1953, 10294.

Mandese, Joe. "In Search of Affluent Viewers." *Television Week* 23(2), January 12, 2004, 47.

The Man Who Shot Liberty Valance, directed by John Ford (John Ford Productions/Paramount Pictures, 1962).

Mann, Klaus. "Cowboy Mentor of the Fuhrer." *Living Age* 359, November 1940, 217–22.

Mar, Gary. "Are Asians Model Minorities?" *Race, Racism and the Law*. University of Dayton, 1993. http://www.academic.udayton.edu/race/01race/model01.htm.

Marcuse, Herbert. *One-Dimensional Man* (Boston: Beacon Press, 1964).

Mark, Eduard. "'Today Has Been a Historical One': Harry S Truman's Diary of the Potsdam Conference." *Diplomatic History* 4(3), Spring 1980, 317–26.

Marling, William, "Detective Fiction on Radio, TV, and in Comics." Detnovel.com. http://www.detnovel.com/OtherMedia.html.

Marples, James. "The Lone Ranger, Freemasonry, and Texas-Ranger Ethics." Alvin Benemerito's Freemasonry Site. http://www.masonic.benemerito.net/msricf/papers/>marples/marples-lone_ranger_freemasonry.pdf.

Marsden, Michael T. "Savior in the Saddle: The Sagebrush Testament." In Jack Nachbar, ed. *Focus on the Western* (Englewood Cliffs, NJ: Prentice-Hall, 1974): 93–100.

Mast, Gerald, and Marshall Cohen. *Film Theory and Criticism* (Oxford: Oxford University Press, 1974).

Matson, Vera, and Lionel Newman. "Theme from *Daniel Boone*" (20th Century Music Corp., 1964).

"May 1958 Staff Notes (2); Abbott Washburn on Spyros Skouras." DDE Diary Series. Box 32. Papers as President (Ann Whitman Files), 1958. Dwight D. Eisenhower Library, Abilene, Kansas.

McArn, Jeff. "Reflections on Political, Religious Uses of the Term "Evil.'" *Observer-Dispatch* (Utica, NY), June 13, 2004. http://collegenews.org/editorials/ 1980/reflections-on-political-religious-uses-of-the-term-evil.html.

McChesney, Robert. *Rich Media, Poor Democracy: Communications Politics in Dubious Times* (New York: The New Press, 1999).

———. *Ruthless Criticism: New Perspectives in US Communication History* (Minneapolis: University of Minnesota Press, 1993).

McGinnis, Joe. *The Selling of the President: The Classic Account of the Packaging of a Candidate* (New York: Penguin Books, 1988).

McNeil, Alex. *Total Television: The Comprehensive Guide to Programming from 1948 to the Present*, 4th ed. (New York: Penguin Books, 1996).

Means, Russell. "Marxism Is a European Tradition." Speech given at Black Hills International Survival Gathering, Rapid City, SD, July 1980. Reprinted in "Fighting Words on the Future of the Earth." *Mother Jones*, December 1980, 26–27.

Means-Coleman, Robin. *African American Viewers and the Black Situation Comedy: Situating Racial Humour* (New York: Garland Books, 2000).

Means-Coleman, Robin, and Emily Yochim. "Symbolic Annihilation." In Wolfgang Donsbach, ed. *The International Encyclopedia of Communication* (Malden, MA: Blackwell Publishing, 2008), 4923.

Memmi, Albert. *The Colonizer and the Colonized* (New York: Orion Press, 1991).

Mercurie. "The Invisible Minority: Native Americans on American Television." *A Shroud of Thoughts* (blog), July 6, 2009. http://mercurie.blogspot.com/2009/07/invisible-minority-native-americans-on.html.

Metcalf, Warren. *Termination's Legacy: The Discarded Indians of Utah* (Lincoln: University of Nebraska Press, 2007).

Metz, Christian. *The Imaginary Signifier: Psycholanalysis and the Cinema* (Bloomington: Indiana University Press 1986; originally published 1977).

Mexal, Stephen J. "Locke, Liberalism, and Western Masculinity in *3:10 to Yuma*." In Jennifer McMahon and Steve Csaki, *The Philosophy of the Western* (Lexington: University Press of Kentucky, 2010), 69–87.

Meyrovitz, Joshua. "Multiple Media Literacies." *Journal of Communications* 48(1), Winter 1998, 96–108.

"Michael Ansara—Perennial Indian." *TV Guide*, April 1960, 29.

Miller, Mark Crispin. "Deride and Conquer." In Todd Gitlin, ed. *Watching Television: A Pantheon Guide to Popular Culture* (New York: Pantheon Books, 1986): 183–228.

———, ed. *Seeing through Movies*: *A Pantheon Guide to Popular Culture* (New York: Pantheon Books, 1990).

Miller, Toby. "Hollywood and the World." In John Hill and Pamela C. Gibson, eds. *The Oxford Guide to Film Studies* (Oxford: Oxford University Press, 1998), 371–81.

Missing in Action, directed by Joseph Zito (Cannon Films/MGM Pictures, 1984).

Mitchell, Lee Clark. *Westerns: Making the Man in Fiction and Film* (Chicago: University of Chicago Press, 1996).

Mittell, Jason. *Genre and Television: From Cop Shows to Cartoons in American Culture* (New York: Routledge, 2004).

The Mod Squad (Thomas-Spelling Productions/Paramount Pictures Television/ABC-TV, 1968–1973).

Moffat, Jay Pierrepont. Diary entry, October 10, 1940. In Nancy H. Hooker, ed. *The Moffat Papers: Selections from the Diplomatic Journals of Jay Pierrepont Moffat, 1919–1943* (Cambridge, MA: Harvard University Press, 1956).

Money, Mary Alice. "Broken Arrows: Images of Native Americans in the Popular Western." In Dane Morrison, ed. *American Indian Studies: An Interdisciplinary Approach to Contemporary Issues* (New York: Peter Lang, 1997): 369–88.

The Monkees (Screen Gems/NBC-TV, 1966–1968).

Montesquieu (Charles de Secondat). *The Spirit of the Laws* (New York: Free Press, 1970; originally published 1748).

Moore, Clayton, with Frank Thomas. *I Was That Masked Man* (New York: Taylor Trade, 1998).

Morrison, Dane, ed. *American Indian Studies: An Interdisciplinary Approach to Contemporary Issues* (New York: Peter Lang, 1997).

Moser, John E. "Gigantic Engines of Propaganda: The 1941 Senate Investigation of Hollywood." *Historian* 63(4), Summer 2001, 731–51.

Mulvey, Laura. "Visual Pleasure and Narrative Cinema." *Screen* 16(3), Autumn 1975, 6–18.

"Museum Archives: FAQ's." Museum of the Cherokee Indian. http://www.cherokeemuseum.org/ html/archives_FAQb.html.

Nachbar, Jack, ed. *Focus on the Western* (Englewood Cliffs, NJ: Prentice-Hall, 1974).

Nachbar, Jack, and Kevin Lause. "All Together Now: Popular Rituals in America." In Jack Nachbar and Kevin Lause, eds. *Popular Culture: An Introductory Text* (Bowling Green, OH: Bowling Green State University Press, 1992), 376–86.

Naficy, Hamid, and Teshome H. Gabriel. *Otherness and the Media: The Ethnography of the Imagined and the Imaged* (Langhorne, PA: Harwood Academic Publishers, 1993).

Naked City (Shelle Productions/Screen Gems/ABC-TV, 1958–1963).

Nakia (series) (David Gerber Productions/Columbia Pictures-Screen Gems/ABC-TV, October 26, 1974).

Nakia, directed by Leonard Horn (David Gerber Productions/Columbia Pictures, April 17, 1974, ABC-TV).

"*Nakia*: Synopsis." MSN Movies. http://movies.msn.com/movies/movie-synopsis/nakia/.

"Nancy Green, the Original Aunt Jemima." African American Registry. http://www.aaregistry.com/african_american_history/1287/Nancy_Green_the_original_Aunt_Jemima.

Nash, Gary, and David Sweet, eds. *Struggle and Survival in Colonial America* (Berkeley: University of California Press, 1981).

"Native Americans." TVAcres.com. http://www.tvacres.com/ethnic_native_american_AK.htm.

Navajo Joe, directed by Sergio Corbucci (Dino De Laurentiis Productions/Twentieth Century-Fox Films, 1967). DVD.

Niose, David, "Noam Chomsky: On Humanism, the Vulnerability of Secular Nationalism, and the Mother of All Book Plugs." *Humanist* 67, January 1, 2007, 20–25.

Nolley, Ken, "The Representation of Conquest: John Ford and the Hollywood Indian (1939–1964)" in Peter C. Rollins and John E. O'Connor, *Hollywood's Indian: The Portrayal of the Native American in Film* (Lexington, KY: University Press of Kentucky, 1998): 73–90.

Noriega, Chon A. "Birth of the Southwest: Social Protest, Tourism, and D. W. Griffith's *Ramona*." In Daniel Bernardi, ed. *The Birth of Whiteness: Race and the Emergence of US Cinema* (New Brunswick, NJ: Rutgers University Press, 1996), 203–26.

Norris, Chuck. "Saving a Roundhouse for Romney." *World Net Daily*, January 7, 2008.

Norris, Chuck, with Ken Abraham. *Against All Odds: My Story* (Nashville, TN: Broadman & Holman, 2006).

Norris, Chuck, with Joe Hyams. *The Secret of Inner Strength: My Story* (Boston: Little, Brown, 1988).

Northern Exposure. Season 3, episode 13, "Things Become Extinct" (Cine-Nevada Productions/Brand-Falsey Productions/MCA Television, January 20, 1992).

Northern Exposure: The Complete First and Second Seasons (Cine-Nevada Productions/Brand-Falsey Productions/MCA Television/Universal Studios, 2006). DVD.

Nudie, Jamie Lee, and Mary Lynn Cabrall. *Nudie: The Rodeo Tailor* (New York: Gibbs Smith, 2004).

Orlando, Joe, and Edward Nelson Bridwell. "TV Scenes We'd Like to See." *Mad* 38, March 1958, 42.

Orwell, George. *Burmese Days* (New York: Mariner Books, 1974; originally published 1934).

Overturf, Sheila. "Program Reviews: *Walker, Texas Ranger*." ChristianAnswers.net, 2002. http://www.christiananswers.net/spotlight/tv/2002/walkertexasranger.html.

Pack, Sam. "Constructing 'the Navajo'": Visual and Literary Representation from Inside and Out." *Wicazo Sa Review* 15(1), Spring 2000, 137–56.

Palast, Greg. "Court Rewards Exxon for Valdez Spill." *Chicago Tribune*, June 26 2008. Reprinted at http://www.gregpalast.com/court-rewards-exxon-for-valdez-oil-spill/?.

Paley, William S. *As It Happened: A Memoir* (New York: Doubleday, 1979).

Parenti, Michael. *Make-Believe Media: The Politics of Entertainment* (New York: St. Martin's Press, 1992; originally published 1986).

Parker, David Wilson. "A Descriptive Analysis of the Lone Ranger as a Form of Popular Art." Unpublished PhD dissertation, Northwestern University, 1956.

Parker, Tyler. "The Horse: Totem Animal of Male Power." In Gerald Mast and Marshall Cohen, eds. *Film Theory and Criticism* (Oxford: Oxford University Press, 1974): 590–99.

Pearce, Roy Harvey. *Savagism and Civilization: A Study of the Indian and the American Mind* (Berkeley: University of California Press, 1964; originally published 1953).

Pearl, Jack. *Garrison's Gorillas* (New York: Dell Books, 1967).

Peary, Gerald, and Jenny Lefcourt. *John Ford: Interviews* (Jackson: University Press of Mississippi, 2001).

Perot, Franceska. "The Re-emergence of the Divine Feminine and Its Significance for Spiritual, Psychological and Evolutionary Growth." Unpublished PhD dissertation, Westbrook University School of Psychology, College of Humanities, 2004.

Phillips, Selene G. "'Indians on Our Warpath': World War II Images of American Indians in *Life* Magazine." In Meta G. Carstarphen and John P. Sanchez, eds. *American Indians in the Mass Media* (Norman: University of Oklahoma Press, 2012), 33–55.

"Pocahontas Legend Hides Unromantic Realities." *Los Angeles Times*, June 24, 1995. http://articles.latimes.com/1995-06-24/local/me-16637_1_historical-pocahontas.

Pollard, Tom. "The Hollywood War Machine." In Carl Boggs, ed. *Masters of War: Militarism and Blowback in the Era of American Empire* (London: Routledge, 2003), 311–42.

"Prayer to the Great Spirit." Peace Winds. http://www.peacewinds.com/main/native/prayer.html.

Presidential Proclamation No. 2914, 15 Federal Regulation 9029, 64 Statute A454. December 16, 1950. Reprinted at *US Code Collection*. Cornell University Law School. http://www.law.cornell.edu/uscode/html/uscode50a/usc_sup_05_50notes.html.

Pumphrey, Martin. "Why Do Cowboys Wear Hats in the Bath? Style Politics for the Older Man." *Critical Quarterly* 31, October 1989, 78–100. Reprinted in Douglas Pye and Ian Cameron, eds. *The Movie Book of the Western* (London: Cassel, 1996): 50–62.

"Puritanism—the Philosophy of the English Revolution." Department of English, Umea University, Umea, Sweden. Accessed May 25, 2011. http://www.eng.umu.se/culturec/PUR.htm.

Pye, Douglas. "Genre and History." In Douglas Pye and Ian Cameron, eds. *The Movie Book of the Western* (London: Cassel, 1996), 9–21.

Pye, Douglas, and Ian Cameron, eds. *The Movie Book of the Western* (London: Cassel, 1996).

"Radio Bill Hearing Set." *New York Times*, February 11, 1951, A40. ProQuest Historical Newspapers.

Rambo: First Blood, Part II, directed by George Cosmatos (Anabasis N.V./Orion Pictures, 1985).

Ramona, directed by David W. Griffith (Biograph Pictures, 1910).

Rand, Yardena. *Wide Open Spaces: Why We Love Westerns* (Manville, RI: Maverick Spirit Press, 2005).

Rango (Thomas-Spelling Productions/ABC-TV, 1967).

Reed, Roy. *Faubus: The Life and Times of an American Prodigal* (Fayetteville: University of Arkansas Press, 1997).

Reynolds, Burt. *My Life* (New York: Hyperion, 1994).

Rodgers, Tom. "Native American Poverty: A Challenge Too Often Ignored." Spotlight on Poverty. September 15, 2009. http://www.spotlightonpoverty.org/ExclusiveCommentary.aspx?id=0fe5c04e-fdbf-4718-980c-0373ba823da7.

Rodriguez, Carlos L., Craig S. Galbraith, and Curt H. Stiles. "American Indian Collectivism: Past Myth, Present Reality." *PERC Reports* 24(2). Property and Environment Research Center. Summer 2006. http://www.perc.org/articles/article802.php.

Roemer, John E., Woojin Lee, and Karine van der Straeten. *Racism, Xenophobia, and Distribution: Multi-Issue Politics in Advanced Democracies* (Cambridge, MA: Harvard University Press, 2007).

Rogin, Michael Paul. "Liberal Society and the Indian Question." *Politics and Society*, May 1971, 269–312. Reprinted in *Ronald Reagan, the Movie and Other Episodes in Political Demonology* (Berkeley: University of California Press, 1987): 134–68.

Roosevelt, Theodore. *The Winning of the West*. Volume 1, *Spread of the English-Speaking Peoples* (New York: G. P. Putnam's Sons, 1910).

Rosaldo, Renato. *Culture and Truth: The Remaking of Social Analysis* (Boston: Beacon Press, 1993).

Rose, Wendy. "The Great Pretenders: Further Reflections on White Shamanism." In M. Annette Jaimes, ed. *The State of Native America: Genocide, Colonization, and Resistance* (Boston: South End Press, 1992), 403.

Rothel, David. *Who Was That Masked Man?* (New York: A. S. Barnes, 1981).

Ryan, Michael, and Douglas Kellner, *Camera Politica: The Politics and Ideology of Contemporary Hollywood Film* (Bloomington: Indiana University Press, 1998).

Sabini, John. *Social Psychology* (New York: W. W. Norton, 1992).

Said, Edward. *Culture and Imperialism* (New York: Vintage Books, 1993).

———. *Orientalism* (New York: Vintage Books, 1979).

Salzberger, Eli M. "Economic Analysis of the Public Domain." In Lucie Guibault and P. Bernt Hugenholtz, eds. *The Future of Public Domain: Identifying the Commons in Information Law* (Alphen aan den Rijn, Netherlands: Kluwer Law International, 2006), 27–59.

Sartre, Jean-Paul. *Being and Nothingness* (New York: Washington Square Press, 1966; originally published 1943).

Saving Grace (Fox Television/Turner Network Television, 2007–2010).

Schatz, Thomas. *Hollywood Genres: Formulas, Filmmaking, and the Studio System* (New York: McGraw-Hill, 1981).

———. "The Structural Influence: New Directions in Film-Genre Study." In Barry Keith Grant, ed. *Film Genre Reader IV* (Austin: University of Texas Press, 2003; originally published 1986), 110–20.

Schiller, Herbert I. *The Mind Managers: How the Master Puppeteers of Politics, Advertising, and Mass Communication Pull the Strings of Public Opinion* (Boston: Beacon Press, 1973).

Schweder, Richard A. "What Do Men Want? A Reading List for the Male Identity Crisis." *New York Times*, January 9, 1994. http://www.nytimes.com/1994/01/09/books/what-do-men-want-a-reading-list-for-the-male-identity-crisis.html.

Scotch, Norman A. "The Vanishing Villains of Television." *Phylon* 21(1), Fall–Winter 1960, 58–62.

The Searchers, directed by John Ford (C. V. Whitney Pictures/Warner Bros. Pictures, 1956).

Seventy-Seven Sunset Strip (Warner Bros. Television/ABC-TV, 1958–1964).

Severo, Richard. "Roy Rogers, Singing Cowboy, Dies at 86." *New York Times*, July 7, 1998, A1.

Shohat, Ella. "Gender and the Culture of Empire." In Hamid Naficy and Teshome Gabriel, eds. *Otherness and the Media: An Ethnography of the Imagined and the Imaged* (Langhorne, PA: Harwood Academic Publishers, 1993), 45–84.

Shostak, Arthur. *Blue-Collar Life* (New York: Random House, 1969).

Shreve, Ivan G., "Premier Collections: The Adventures of Philip Marlowe, Volume One" (liner notes). *Radio Archives.Org*. http://www.radioarchives.org/sets/PC29.htm.

Siegel, Jerry, and Joe Shuster. *Action Comics* 1 (National Periodicals), June 1938. Excerpted at "Religion of Superman/Clark Kent." Comic Book Religion. http://www.comicbookreligion.com/?c=1&Superman_Clark_Kent.

Siegel, Robert. "The Lone Ranger: Justice from Outside the Law." National Public Radio. January 17, 2008. http://www.npr.org/templates/story/story.php?storyId=18195323.

Simpson, Christopher. *Science of Coercion: Communication Research and Psychological Warfare, 1945–1960* (Oxford: Oxford University Press, 1994).

Sinfield, Alan. *Faultlines: Cultural Materialism and the Politics of Dissident Reading* (Berkeley: University of California Press, 1992).

"Sixkiller Assassinated: Cherokee Chief of Police Riddled with Bullets." *New York Times*, December 27, 1886, 1. ProQuest Historical Newspapers.

Sloan, William, and David Startt. *The Media in America: A History* (Northport, AL: Vision Press, 1999).

Slotkin, Richard. *The Fatal Environment: The Myth of the Frontier in the Age of Industrialization, 1800–1890* (Norman: University of Oklahoma Press, 1998).

———. *Regeneration through Violence: The Mythology of the American Frontier, 1600–1860* (Norman, OK: University of Oklahoma Press, 1973), 357.

Smith, Philip D., and Kristin Natalier. *Understanding Criminal Justice: A Sociological Perspective* (Thousand Oaks, CA: Sage Publications, 2004).

Smith, Robert W., and Elisabeth A. Smith, eds. *Scriptural and Secular Prophecies Pertaining to the Last Days Including Prophecies by Members of the Church of Jesus Christ of Latter Day Saints* (Whitefish, MT: Kessinger Publishing, 2004).

Smith, William C. "God's Plan in America." *New Age Magazine*, 58(9), September 1950, 551. Reprinted in Phillip D. Collins and Paul David Collins. *The Ascendancy of the Scientific Dictatorship: An Examination of Epistemic Autocracy from the Nineteenth to the Twenty-first Century* (Bloomington, IN: iUniverse, 2004), 73.

Soldier Blue, directed by Ralph Nelson (Katzka-Loeb Productions/Avco-Embassy Pictures/Lionsgate 2006; originally released 1970).

Solomon, Norman. "The Military-Industrial-Media Complex: Why War Is Covered from the Warriors' Perspective." Fairness and Accuracy in Reporting. August 2005. http://www.fair.org/index.php?page=2627.

Southern, Joe, "The Lone Ranger a True Believer," *The Silver Bullet*, n.d. http://www.series.net/ranger/index/html. Retrieved February 20, 2008.

Sowell, Thomas. *Race and Culture: A World View* (New York: Basic Books, 1994).

Span, Paula. "That Lone Star Good Guy: George Bush? Nope. Walker, Texas Ranger." *Washington Post*, May 19, 2001, C1.

Spenser, Edmund. "Sonnett XIV." In Robert Atwan and Laurance Wieder, eds. *Chapters into Verse: A Selection of Poetry in English Inspired by the Bible from Genesis through Revelation* (Oxford: Oxford University Press, 2000), 448.

"The Spider Women." *Time*, September 11, 1972. http://www.time.com/time/magazine/article/0,9171,906347,00.html.

Spivak, Gayatri Chakravorty. "Can the Subaltern Speak?" In Cary Nelson and Lawrence Grossberg, eds. *Marxism and the Interpretation of Culture* (Champaign: University of Illinois Press, 1988), 271–313.

Stagecoach, directed by John Ford (Walter Wanger Productions/United Artists, 1939).

Stam, Robert, and Louise Spence. "Colonialism, Racism, and Representation." In Bill Nichols, ed. *Movies and Methods: An Anthology*. Volume 2 (Berkeley: University of California Press, 1985), 632–49.

Stannard, David. *American Holocaust: The Conquest of the New World* (Oxford: Oxford University Press, 1992).

Star Trek (Desilu Productions/NBC-TV, 1966–1969).

Stedman, Raymond William. *Shadows of the Indian* (Norman: University of Oklahoma Press, 1982).

Stern, Matthew A. “Review: The Best of *Bravestarr.*” PopMatters. July 18, 2007. http://www.popmatters.com/pm/review/the-best-of-bravestar.

Stiffarm, Lenore A., and Phil Lane Jr. “The Demography of North America: A Question of American Indian Survival.” In M. Annette Jaimes, ed. *The State of North America: Genocide, Colonization, and Resistance* (Boston: South End Press, 1992), 23–54.

Storey, John, ed. *Cultural Theory and Popular Culture: A Reader* (Athens: University of Georgia Press, 1998; originally published 1968).

Sugrue, Thomas. “A Dream Still Deferred.” *New York Times* (editorial), March 26, 2011. http://www.nytimes.com/2011/03/27/opinion/27Sugrue.html.

Suid, Lawrence. *Guts and Glory: The Making of the American Military Image in Film* (Lexington: University Press of Kentucky, 2002).

———. *Sailing on the Silver Screen: Hollywood and the U.S. Navy* (Annapolis, MD: Naval Institute Press, 1996).

Swenson, Allan A., and Michael Benson. *The Complete Idiot's Guide to the CIA* (New York: Alpha Books, 2002).

Tamarkin, Mordechai. *Cecil Rhodes and the Cape Afrikaaners: The Imperial Colossus and the Colonial Parish Pump* (London: Frank Cass, 1996).

Taylor, Annette M. “Cultural Heritage in *Northern Exposure*.” In S. Elizabeth Bird, ed. *Dressing in Feathers: The Construction of the American Indian in American Popular Culture* (Boulder, CO: Westview Press, 1996), 229–44.

Teeple, Donald Shea. “TV Westerns Tell a Story.” *American Mercury*, April 1958, 115–17.

Telotte, J. P. “A Fate Worse Than Death.” *Journal of Popular Film and Television* 26(3), 1998, 120–27.

Thank You, Mask Man, directed by Lenny Bruce and John Magnusson (Columbus Productions, 1968). Posted on YouTube December 3, 2005. http://www.youtube.com/watch?v=NZMTPGc0Z6A.

Thernstrom, Stephan, and Abigail Thernstrom. *America in Black and White: One Nation, Indivisible* (New York: Touchstone, 1997).

Thompson, Chuck. “Great Movie Indians.” *Cowboys & Indians*, June 2011, 67.

Tindall, George Brown, *America: A Narrative History* (New York: W. W. Norton, 1992; originally published 1984), 57.

Todorov, Tzvetan. *The Conquest of America: The Question of the Other* (New York: Harper and Row, 1984).

Tonight: Four Decades of “The Tonight Show” starring Johnny Carson (Respond 2 Entertainment, 2010). DVD.

Torgovnick, Marianna. *The War Complex: World War II in Our Time* (Chicago: University of Chicago Press, 2005).

Torres, Sasha, ed. *Living Color: Race and Television in the United States* (Durham, NC: Duke University Press, 1998).

Trendle, George W. Letter to Ines Wallace. December 12, 1956. George Trendle papers, Burton Collection. Detroit Public Library.

Trevathan, Kim. *Coldhearted River: An Oddyssey down the Cumberland* (Knoxville: University of Tennessee Press, 2006).

Tucker, Frank H. *The White Conscience: A Hard-Hitting, Documented Survey of 500 Years of Western Domination* (New York: Frederick Ungar, 1968).

Turner, Graeme. *Film as Social Practice* (London: Routledge, 2000).

Tuveson, Ernest Lee. *Redeemer Nation: The Idea of America's Millennial Role* (Chicago: University of Chicago Press, 1968).

Twentieth Century-Fox Hour. Episode 15: “Broken Arrow” (Twentieth Century-Fox Pictures/CBS-Television, May 2, 1956).

Twin Peaks: The First Season (Lynch-Frost Productions/Republic Pictures/Artisan Entertainment, 2001). DVD.

Twin Peaks: The Second Season (Lynch-Frost Productions/CBS-Paramount Video, 2007). DVD.

The Untouchables (Langford Productions/Desilu Productions/ABC-TV, 1959–1963).

"Urge CBS, ABC, NBC-TV to Treat Negroes Fairly." *Jet* 22(10), June 28 1962, 62.

U.S. Department of State Bulletin. May 20, 1963, 783. Cited in Frank H. Tucker, *The White Conscience: A Hard-Hitting, Documented Survey of 500 Years of Western Domination* (New York: Frederick Ungar, 1968), 334, fn 11.

Van Alstyne, Richard W. *The Rising American Empire* (New York: W. W. Norton, 1960).

Vera, Hernan, and Andrew Gordon. *Screen Saviors: Hollywood Fictions of Whiteness* (Lanhan, MD: Rowman & Littlefield, 2003).

———. "Sincere Fictions of the White Self in the American Cinema: The Divided White Self in Civil War Films." In Daniel Bernardi, Classic Hollywood, Classic Whitness (Minneapolis: University of Minnesota Press, 2001), 263–80.

Wagon Train (Revue Studios/NBC-TV, 1957–1965).

Walker, Texas Ranger: Trial by Fire (Norris Brothers Entertainment/Paramount Pictures/CBS Television, October 21, 2005).

Walker, Texas Ranger: The Complete First Season (Paramount/CBS Home Video, 2006). DVD.

Walker, Texas Ranger: The Complete Second Season (Paramount/CBS Home Video, 2007). DVD.

Walker, Texas Ranger: The Complete Fourth Season (Paramount/CBS Home Video, 2007). DVD.

Walker, Texas Ranger: The Complete Sixth Season (Paramount/CBS Home Video, 2007). DVD.

Watson, Mary Ann. *The Expanding Vista: American Television in the Kennedy Years* (Durham, NC: Duke University Press, 1994).

Weber, Max. *The Protestant Ethic and the Spirit of Capitalism* (New York: W. W. Norton, 2009; originally published 1905).

Wexman, Virginia Wright. "The Family on the Land: Race and Nationhood in Silent Westerns." In Daniel Bernardi, ed. *The Birth of Whiteness: Race and the Emergence of US Cinema* (New Brunswick, NJ: Rutgers University Press, 1996), 129–69.

White, Hayden. *Figural Realism: Studies in the Mimesis Effect* (Baltimore: Johns Hopkins University Press, 1999).

Wiegman, Robyn. "Race, Ethnicity, and Film." In John Hill and Pamela Church Gibson, eds. *The Oxford Guide to Film Studies* (Oxford: Oxford University Press, 1998), 158–68.

Wiggershaus, Rolf. *The Frankfurt School: Its History, Theories, and Political Significance* (Boston: MIT Press, 1995; originally published 1986).

Wilson, Pamela. "Confronting the 'Indian Problem': Media Discourses of Race, Ethnicity, Nation, and Empire in 1950s America." In Sasha Torres, ed. *Living Color: Race and Television in the United States* (Durham, NC: Duke University Press, 1998), 35–61.

Wirbel, Loring. "Star Wars: Imperialism in Space." In Carl Boggs, ed. *Masters of War: Militarism and Blowback in the Era of American Empire* (London: Routledge, 2003), 59–82.

Wood, Robin. *Hollywood from Vietnam to Reagan* (New York: Columbia University Press, 1986).

Woodford, Frank B. *Lewis Cass: The Last Jeffersonian* (New York: Octagon Books, 1973).

"The World of Harry Truman." *Time*, January 8, 1973. http://www.time.com/time/ magazine/article/0,9171,910501-4,00.html.

Wright, Will. *Six Guns and Society* (Berkeley: University of California Press, 1977).

Wu, Frank. *Race in America beyond Black and White* (New York: Basic Books, 2002).

Yeshin, Tony. *Advertising* (Florence, KY: Cengage Learning Business Press, 2005).

Zackel, Frederick, "Robinson Crusoe and the Ethnic Sidekick." *Bright Lights Film Journal* 58, November 2007. http://www.brightlightsfilm.com/30/crusoe1.html.

Zizek, Slavoj. *The Ticklish Subject: The Absent Centre of Political Ontology* (London: Verso, 1999).

Index

About the Author

Michael Ray FitzGerald is a communication scholar and freelance journalist. He holds a PhD in media studies from the University of Reading (United Kingdom) and a master's degree in mass communication from the University of Florida. His focus is media history; his research specialties are minorities in mass media, ideology in mass media, and Cold War television. His work has been published in *Howard Journal of Communications*, *Journal of Popular Television*, *Film & History*, *Southern Cultures*, *American Indian Culture and Research Journal*, *Utne Reader*, and others. He is the author of a collection titled *Mixed Metaphors*. FitzGerald lives in Jacksonville, Florida, and teaches communications at Jacksonville University and College of Coastal Georgia in Kingsland. His website is located at michaelrayfitzgerald.com.